T0245094

In 1993, Professor Oleinik was invited to give a series of lectures about her work in the area of partial differential equations. This book contains those lectures, and more. It is in two parts, the first being devoted to the study of the asymtotic behaviour at infinity of solutions of a class of non-linear second order elliptic equations in unbounded, in particular mathematical physics, such as in the theory of traveling waves, homogenization, boundary layer theory, flame propagation and combustion.

The second part contains the most recent results of the author's research in the theory of homogenization of partial differential equations, and is concerned with questions about partially perforated domains and of solutions with rapidly alternating types of boundary conditions. These new asymptotic problems arise naturally in applications.

Many of the results within have not appeared in book form before, and this volume sheds new light on the subject, raising many new ideas and open problems.

Some asymptotic problems in the theory of partial differential equations

Lezioni Lincee
Editor: Luigi A. Radicati di Brozolo,
Scuola Normale Superiore, Pisa

This series of books arises from lectures given under the auspices of the Accademia Nazionale dei Lincei and is sponsored by *Foundazione IBM Italia.*

The lectures, given by international authorities, will range on scientific topics from mathematics and physics through to biology and economics. The books are intended for a broad audience of graduate students and faculty members, and are meant to provide a *'mise au point'* for the subject they deal with.

The symbol of the Accademia, the lynx, is noted for its sharp sightedness; the volumes in the series will be penetrating studies of scientific topics of contemporary interest.

Already published

Chaotic Evolution and Strange Attractors: D. Ruelle
Introduction to Polymer Dynamics: P. de Gennes
The Geometry and Physics of Knots: M. Atiyah
Attractors for Semigroups and Evolution Equations:
    O. Ladyzhenskaya
Half a century of Free Radical Chemistry:
    D.N.R. Barton in collaboration with S.I. Parekh
Asymptotic Behaviour of Solutions of Evolutionary Equations:
    M.I. Vishik
Bound Carbohydrates in Nature: L. Warren
Neural Activity and the Growth of the Brain: D. Purves
Perspectives in Astrophysical Cosmology: M. Rees

Some asymptotic problems in the theory of partial differential equations

OLGA OLEINIK
Faculty of Mechanics and Mathematics, Moscow State University

CAMBRIDGE
UNIVERSITY PRESS

CAMBRIDGE UNIVERSITY PRESS
Cambridge, New York, Melbourne, Madrid, Cape Town, Singapore, São Paulo

Cambridge University Press
The Edinburgh Building, Cambridge CB2 8RU, UK

Published in the United States of America by Cambridge University Press, New York

www.cambridge.org
Information on this title: www.cambridge.org/9780521480833

First published 1996

A catalogue record for this publication is available from the British Library

ISBN 978-0-521-48083-3 hardback
ISBN 978-0-521-48537-1 paperback

Transferred to digital printing 2007

Contents

Preface

This book is based on lectures delivered by the author at the invitation of the Committee of Accademia Nazionale dei Lincei and its President, Professor G. Salvini.

The book consists of two parts and reflects the scientific interests of the author during the last two years. Most of the results presented here are new and are being published for the first time.

The first part of the book is devoted to the study of the asymptotic behavior at infinity of solutions of a class of nonlinear second order elliptic equations in unbounded domains and, in particular, in cylindrical domains. The model equation for this class may be $\Delta u - |u|^{p-1}u = 0$, $p = \text{const} > 1$. Questions of this kind occur in many problems of mathematical physics, in the theory of travelling waves, homogenization, stationary states, boundary layer theory, biology, flame propagation, the probability theory and so on.

The main problems treated in the first part are the study of solutions of the boundary value problems with Dirichlet or Neumann conditions on the lateral part of the boundary of a cylindrical domain and to find the asymptotics of the solutions at infinity.

Specifically, the Gauss equation $\Delta u - e^u = 0$ is considered in one section of the book and the asymptotic properties at infinity of its solutions in cylindrical domains are studied.

Also, the asymptotic behavior of solutions of boundary-value problems in a neighbourhood of a conic point for nonlinear second order equations is investigated. Some new phenomena are found which do not appear in the case of linear equations.

The second part of the book, which contains the most recent results of the author in the theory of homogenization of partial differential equations, is concerned with homogenization in partially perforated domains and homogenization of solutions with rapidly alternating types of boundary conditions.

The problems of homogenization were studied by the author for many years and two of her books on the subject have recently appeared: *Mathematical problems in elasticity and homogenization* (with A.S. Shamaev and G.A. Yosifian, North-Holland, Amsterdam, 1992) and *Homogenization of differential operators and integral functionals* (with V.V. Jikov and S.M. Kozlov, Springer-Verlag, 1994). In these books one can find many basic results on homogenization of partial differential equations as well as the main techniques and methods used in this theory.

The present book raises many interesting open problems some of which are only just beginning to be studied.

The author would like to express her cordial thanks to Professor G. Salvini for his kind invitation, to Professor Gaetano Fichera and Professor G. Kallianpur for their helpful remarks and to Matelda Fichera without whose invaluable help, during the author's stay in Rome, this book could not have been completed.

Olga Oleinik
Rome, 27 October 1993.

Chapter 1

Asymptotic problems for nonlinear elliptic equations

1.1 Nonlinear elliptic boundary-value problems in unbounded domains and the asymptotic behavior of their solutions.

In this section we consider the problem of existence, uniqueness and asymptotic properties at infinity of solutions of boundary-value problems in unbounded domains for a class of nonlinear second order elliptic equations of the form

$$\sum_{i,j=1}^{n} \frac{\partial}{\partial x_i} \left(a_{ij} \frac{\partial u}{\partial x_j} \right) - a(x)|u|^{p-1}u = f(x). \tag{1}$$

Equations of this type were studied in many papers (see, for example, [1]–[4]), [9]–[27].

Let Ω be a smooth unbounded domain in $\mathbf{R}^n = (x_1, \ldots, x_n)$. In particular, Ω can be all of $\mathbf{R}^n$. We consider the boundary-value problem for the equation (1) in Ω, with the boundary conditions

$$u = 0 \text{ on } \Gamma_D, \qquad \frac{\partial u}{\partial \mu} = 0 \text{ on } \Gamma_N, \tag{2}$$

where $\partial\Omega = \Gamma_D \cup \Gamma_N$, and Γ_D or Γ_N can be the empty set on $\partial\Omega$. We assume that $a_{ij} \in L^\infty_{\text{loc}}(\Omega)$, and

$$\sum_{i,j=1}^{n} a_{ij}(x)\xi_i\xi_j \geqslant \lambda|\xi|^2, \qquad x \in \Omega, \ \xi \in \mathbf{R}^n,$$

1

$\lambda = const > 0$, $p > 1$, $a \in L^1_{\text{loc}}(\Omega)$ and $a(x) \geqslant a_0 = const > 0$, for any $x \in \Omega$. Here as usual

$$\frac{\partial u}{\partial \mu} \equiv \sum_{i,j=1}^{n} a_{ij}(x) \frac{\partial u}{\partial x_j} \nu_j,$$

where $\nu = (\nu_1, \ldots, \nu_n)$ is the outward unit normal vector to $\partial \Omega$.

We introduce the following notations:

$$B_R = \{x \,:\, |x| < R\}, \ \Omega_R = \Omega \cap B_R,$$
$$V(\Omega_R) = \{w \,:\, w \in H^1(\Omega_R), \ w = 0 \text{ on } \Gamma_D \cap B_R\},$$
$$V_{\text{loc}}(\Omega) = \{w \,:\, w \in V(\Omega_R) \text{ for any } R > 0\}.$$

We also consider the dual spaces $V^*(\Omega_R)$ and $V^*_{\text{loc}}(\Omega)$ respectively. Let $f \in V^*_{\text{loc}}(\Omega)$. The space $H^1(G)$ is a Hilbert space with the scalar product $\int_G (uv + \nabla u \cdot \nabla v) dx$.

A function $u \in V_{\text{loc}}(\Omega)$ is called a weak solution of problem (1),(2), if $a(x)|u|^{p-1}u \in L^1_{\text{loc}}(\Omega)$ and

$$\sum_{i,j=1}^{n} \int_{\Omega} a_{ij} \frac{\partial u}{\partial x_j} \frac{\partial \varphi}{\partial x_i} dx + \int_{\Omega} a(x)|u|^{p-1}u\varphi dx = - <f, \varphi> \qquad (3)$$

for any $\varphi \in V_{\text{loc}}(\Omega) \cap L^\infty_{\text{loc}}(\Omega)$, which has a compact support in $\Omega \cup \Gamma_N$.

Let us introduce the cut-off function

$$\Theta_R(x) = \theta^2 \left(\frac{|x|}{R}\right) \ \text{ for } R > 0,$$

where $\theta \in C^\infty(\mathbf{R})$ is such that $\theta(s) = 1$ if $|s| \leqslant 1/2$ and $\theta(s) = 0$ for $|s| \geqslant 1$.

Theorem 1. *Assume that for* $u \in V(\Omega_N)$ *and* $a(x)|u|^{p+1} \in L^1(\Omega_N)$ *the inequality*

$$\sum_{i,j=1}^{n} \int_{\Omega_N} a_{ij} \frac{\partial u}{\partial x_j} \frac{\partial(u\Theta_R)}{\partial x_i} dx + \int_{\Omega_N} a(x)|u|^{p+1}\Theta_R dx \leqslant - <f, u\Theta_R> \qquad (4)$$

holds for some $f \in V^*(\Omega_R)$ *and any* $R \leqslant N$. *Then there exist a constant* $C > 0$ *independent of* Ω, R *and* u *such that if* $R \in [1, N]$ *we have*

$$\int_{\Omega_{R/2}} |\nabla u|^2 dx \leqslant R^{n-2-4/(p-1)} \left(C + R^{(4p-n(p-1))/(p-1)} \|f\|^2_{V^*(\Omega_R)}\right) \qquad (5)$$

and

$$\int_{\Omega_{R/2}} |u|^{p+1} dx \leqslant R^{n-2(p+1)/(p-1)} \left(C + R^{(4p-n(p-1))/(p-1)} \|f\|^2_{V^\bullet(\Omega_R)} \right).$$
$$(6)$$

Proof. Let $R = 1$. From (4) and the assumptions $p > 1$, $a(x) \geqslant a_0 = const > 0$ it follows that

$$C_1 \int_{\Omega_1} |\nabla u|^2 \theta^2(|x|) dx + a_0 \int_{\Omega_1} |u|^{p+1} \theta^2(|x|) dx$$
$$\leqslant 2\epsilon \int_{\Omega_1} |\nabla u|^2 \theta^2(|x|) dx + C_2 \int_{\Omega_1} |u|^2 |\theta'(|x|)|^2 dx$$
$$+ |<f, u\theta^2>|, \qquad C_1, C_2 = const. \qquad (7)$$

Using the Young inequality $\left(ab \leqslant \dfrac{a^s}{s} + \dfrac{b^q}{q}, \dfrac{1}{s} + \dfrac{1}{q} = 1 \right)$, we obtain from (7) that

$$\int_{\Omega_1} |\nabla u|^2 \theta^2(|x|) dx + a_0 \int_{\Omega_1} |u|^{p+1} \theta^2(|x|) dx$$
$$\leqslant C_3 \int_{\Omega_1} |u|^2 \theta^K \theta^{-K} |\theta'|^2 dx + |<f, u\theta^2>|$$
$$\leqslant \delta \int_{\Omega_1} |u|^{p+1} \theta^2 dx + C_4 \int_{\Omega_1} |\theta'|^{2(p+1)/(p-1)} \theta^{-4/(p-1)} dx$$
$$+ |<f, u\theta^2>|,$$
$$s = (p+1)/2, \ q = (p+1)/(p-1), K = 2s. \qquad (8)$$

By the definition of f

$$|<f, u\theta^2>| \ \leqslant \ C_5 \|f\|^2_{V^\bullet(\Omega_1)} + \delta_1 \|u\theta^2\|^2_{H^1(\Omega_1)}$$
$$\leqslant \ C_5 \|f\|^2_{V^\bullet(\Omega_1)}$$
$$+ \delta_2 \left(\int_{\Omega_1} u^2 \theta^4 dx + \int_{\Omega_1} |\nabla u|^2 \theta^2 dx + \int_{\Omega_1} u^2 |\theta'|^2 dx \right). (9)$$

The last integral is already considered above (see (8)). Applying the Young inequality again we obtain

$$\int_{\Omega_1} u^2\theta^4 dx \leqslant \delta_3 \int_{\Omega_1} u^{p+1}\theta^2 dx + C_6 \int_{\Omega_1} \theta^{2+2\frac{(p+1)}{(p-1)}} dx. \tag{10}$$

Here ϵ, δ, δ_1, δ_2 are arbitrarily small numbers. From (7)–(10) we have

$$\int_{\Omega_1} |\nabla u|^2\theta^2(|x|)dx + \int_{\Omega_1} |u|^{p+1}\theta^2(|x|)dx$$

$$\leqslant C_7\|f\|^2_{V^*(\Omega_1)} + C_8 \int_{\Omega_1} |\theta'|^{2(p+1)/(p-1)}\theta^{-4/(p-1)}dx$$

$$+C_9 \int_{\Omega_1} \theta^{4p/(p-1)}dx, \quad C_j = const. \tag{11}$$

Let us choose $\theta(s)$ such that $\theta'(s) = \mathcal{O}\left((1-s)^{m-1}\right)$ as $s \to 1$. If $m > (p+1)/(p-1)$, then the second integral is finite and we have

$$\int_{\Omega_1} |\nabla u|^2\theta^2 dx + \int_{\Omega_1} |u|^{p+1}\theta^2 dx \leqslant C_{10} + C_7\|f\|^2_{V^*(\Omega_1)}. \tag{12}$$

The last inequality yields us the inequality (5),(6) for $R = 1$. In order to get the inequality (5),(6) for $R > 1$ we introduce in (4) the change of variables

$$x' = \frac{x}{R}, \quad u(x) = R^{-2/(p-1)}v(Rx'). \tag{13}$$

In the new variables x', v we have (5),(6) for $R = 1$. If we write the inequalities thus obtained in terms of the variables x, u, we obtain (5),(6) for any $R \leqslant N$.

We use Theorem 1 to prove the existence and uniqueness results for the problem (1),(2).

Theorem 2. *Under the above conditions on the data of the problem (1),(2) there exists a unique weak solution $u(x)$ of problem (1),(2) such that $a(x)|u|^{p+1} \in L^1_{loc}(\Omega)$ and the integral identity (3) holds also for $\varphi = u\theta^2\left(\frac{|x|}{R}\right)$ for any $R \geqslant 1$.*

Proof. (Existence) In order to construct a solution of the problem (1),(2) we consider the following boundary-value problem:

$$\sum_{i,j=1}^{n} \frac{\partial}{\partial x_i}\left(a_{ij}\frac{\partial u_N}{\partial x_j}\right) - a(x)|u_N|^{p-1}u_N = f(x) \text{ in } \Omega_N, \qquad (14)$$

$$\frac{\partial u_N}{\partial \mu} = 0 \text{ on } \Gamma_N \cap B_N; \quad u_N = 0 \text{ on } (\Gamma_D \cap B_N) \cup \{x \in \Omega : |x| = N\}. \quad (15)$$

The existence and uniqueness of a solution $u_N \in H^1(\Omega_N)$ satisfying (14),(15) in the sense that $a(x)|u_N|^p \in L^1(\Omega_N)$ and the integral identity (3) holds with Ω replaced by Ω_N and for any $\varphi \in H^1(\Omega_N) \cap L^\infty(\Omega_N)$ with $\varphi = 0$ on $(\Gamma_D \cap B_N) \cup \{x \in \Omega : |x| = N\}$ is a consequence of the results of H. Brezis and F. Browder [5]. Their results also imply that $a(x)|u_N|^{p+1} \in L^1(\Omega_N)$ and that the integral identity also holds for $\varphi = u_N$. In fact, if $0 < R < N$ we have

$$\sum_{i,j=1}^{n}\int_{\Omega_R} a_{ij}\frac{\partial u_N}{\partial x_j}\frac{\partial(u_N\theta^2)}{\partial x_i}\mathrm{d}x + \int_{\Omega_R} a(x)|u_N|^{p+1}\theta^2\mathrm{d}x = \langle f, u_N\Theta^2\rangle \qquad (16)$$

In order to prove (16) we take $\varphi = T_m(u_N)\theta^2$ in the integral identity associated to (3) but on Ω_N, where $T(s) = \min(m, |s|)\text{sign } s$. Making $m \to \infty$ we obtain (16). By Theorem 1 we have

$$\int_{\Omega_{R/2}} \left(|\nabla u_N|^2 + |u_N|^{p+1}\right)\mathrm{d}x \leqslant C$$

for some constant C depending on R, but independent of N. By standard results we get that $\{u_N\}$ is bounded in $H^1(\Omega_R)$ and by diagonal extraction it follows that there exist u and a sub-sequence u_N such that $u_N \to u$ in $H^1_{\mathrm{loc}}(\Omega)$ weakly, in $L^{p+1}_{\mathrm{loc}}(\Omega)$ weakly, in $L^2_{\mathrm{loc}}(\Omega)$ strongly and almost everywhere in Ω. Passing to the limit as $N \to \infty$ in the integral identity for u_N we obtain (3).

(Uniqueness): Let us prove that the weak solution $u(x)$ of the problem (1),(2) constructed above is unique. Let u_1, u_2 be weak solutions of the problem (1),(2) such that $a(x)|u_i|^{p+1} \in L^1_{\mathrm{loc}}(\Omega)$, and the integral identity (3) holds for $\varphi = u_i\Theta_R$ for any $R > 1$, $i = 1, 2$. Let $v = u_1 - u_2$. Since $v \in V_{\mathrm{loc}}(\Omega) \cap L^{p+1}_{\mathrm{loc}}(\Omega)$, arguing as in the proof of (16), we can take $\varphi = v\Theta_R$ in the integral identity (3) associated to u_i. We get

$$\int_{\Omega_R}\sum_{i,j=1}^{n} a_{ij}\frac{\partial v}{\partial x_j}\frac{\partial(v\Theta_R)}{\partial x_i}\mathrm{d}x + \int_{\Omega_R} A(x)|v|^{p+1}\Theta_R\mathrm{d}x = 0, \qquad (17)$$

where

$$A(x) \equiv \begin{cases} \dfrac{a(x)\left[|u_1(x)|^{p-1}u_1(x) - |u_2(x)|^{p-1}u_2(x)\right]}{|u_1(x) - u_2(x)|^{p-1}(u_1(x) - u_2(x))}, & \text{if } u_1(x) \neq u_2(x), \\ a_0, & \text{if } u_1(x) = u_2(x). \end{cases}$$

It is easy to prove by considering the function

$$\frac{1 - |y|^{p-1}y}{(1-y)|1-y|^{p-1}},$$

that $A(x) \geqslant A_0 = const > 0$. Therefore, Theorem 1 can be applied to the function v with $f \equiv 0$. According to Theorem 1 we have

$$\int_{\Omega_{R/2}} |\nabla v|^2 dx \leqslant CR^{n-2-4/(p-1)}, \tag{18}$$

$$\int_{\Omega_{R/2}} |v|^{p+1} dx \leqslant CR^{n-2(p+1)/(p-1)}, \qquad C = const. \tag{19}$$

If $1 < p < (n+2)/(n-2)$ or $n = 2$, then from (19) it follows that $v = 0$. If $p \geqslant (n+2)/(n-2)$, we shall prove that $v = 0$ in Ω by contradiction. Suppose that there exists $\alpha > 0$ such that $v(x) \geqslant \alpha$ on a set $\omega \subset \Omega$ and meas $\omega > 0$. It is easy to prove that the function $w = (v - \alpha)_+$, where $(v - \alpha)_+ = v - \alpha$ for $v - \alpha > 0$ and $(v - \alpha)_+ = 0$ for $v - \alpha \leqslant 0$, satisfies the equality

$$\int_{\Omega} \sum_{i,j=1}^{n} a_{ij} \frac{\partial w}{\partial x_j} \frac{\partial(w\Theta_R)}{\partial x_i} dx + \int_{\Omega} A(x) \frac{(w+\alpha)^{p+1}}{w^{q+1}} w^{q+1}\Theta_R dx = 0, \tag{20}$$

where $q < (n+2)/(n-2)$. Since $(w+\alpha)^{p+1}/w^{q+1} \geqslant b > 0$ for any $w > 0$ and $q < p$, where $b = const > 0$, we can apply Theorem 1 to w. We then have

$$\int_{\Omega_{R/2}} |w|^{q+1} dx \leqslant CR^{n-2(q+1)/(q-1)}. \tag{21}$$

From (21) it follows that $w \equiv 0$ in Ω. That contradicts the assumption that $w > 0$ on $\omega \subset \Omega$ with meas $\omega > 0$.

Remark 1. The first result where the existence and uniqueness of a solution of the boundary-value problem (1),(2) with condition $\Gamma_N = \emptyset$ were proved without the growth conditions at infinity on f and u was by H.

Brezis [2]. The H. Brezis uniqueness result uses an explicit supersolution and his method cannot be applied to the case $\Gamma_N \neq \emptyset$.

Consider now the behavior as $|x| \to \infty$ of a weak solution of problem (1),(2). We study the case $f = 0$ in $\Omega \backslash B_{R'}$, where R' is some constant.
We consider the function $u \in H^1_{\text{loc}}(\Omega \backslash B_{R'})$ such that

$$a(x)|u|^{p+1} \in L^1_{\text{loc}}(\Omega \backslash B_{R'}),$$

$u \in L^\infty(\Omega \cap \partial B_{R'})$ and satisfies the equation

$$\sum_{i,j=1}^{n} \frac{\partial}{\partial x_i} \left(a_{ij}(x) \frac{\partial u}{\partial x_j} \right) - a(x)|u|^{p-1}u = f(x) \quad \text{in } D'(\Omega \backslash B_{R'}). \qquad (22)$$

Theorem 3. *Assume that* $u = 0$ *on* $\partial\Omega \backslash B_{R'}$, $a_{ij}(x) = const.$, $i, j = 1, \ldots, n$. *Then*

$$|u(x)| \leqslant C|x|^{-2/(p-1)} \quad \text{in } \Omega \backslash B_{R'}, \quad C = const > 0. \qquad (23)$$

Proof. Let $N > R$. We set $\Omega_N^{R'} = \{x \in \Omega : R' < |x| < N\}$. Let v_N be a weak solution of the problem

$$\sum_{i,j=1}^{n} \frac{\partial}{\partial x_i} \left(a_{ij}(x) \frac{\partial v_N}{\partial x_j} \right) - a(x)|v_N|^{p-1}v_N = 0 \quad \text{in } \Omega_N^{R'} \qquad (24)$$

with the boundary conditions

$$v_N = u \text{ on } \overline{\Omega} \cap \partial B_{R'}, \quad \text{and} \quad v_N = 0 \text{ on } (\partial\Omega \cap (B_N \backslash B_{R'}) \cup (\Omega \cap \partial B_N)). \qquad (25)$$

Consider the function $V(x) = K|x|^{-2/(p-1)}$. It is easy to check that for $a_{ij}(x) = const.$

$$\sum_{i,j=1}^{n} \frac{\partial}{\partial x_i} \left(a_{ij}(x) \frac{\partial V}{\partial x_j} \right) - a(x)|V|^{p-1}V \leqslant 0$$

and $V(x) \geqslant v_N(x)$ on $\Omega \cap \partial B_{R'}$, if $K > 0$ is a sufficiently large constant. Applying the maximum principle, we deduce that

$$|v_N(x)| \leqslant K|x|^{-2/(p-1)} \quad \text{in } x \in \Omega_N^{R'}. \qquad (26)$$

Letting $N \to \infty$ we obtain (23), the assertion of the theorem.

The following result shows that the geometry of Ω can have influence on the decay of $u(x)$ as $|x| \to \infty$.

Theorem 4. *Assume that*

$$diameter(\Omega \cap \{x : |x| = N\}) \leqslant T, \quad \forall N \geqslant R', \qquad (27)$$

where the constant T does not depend on N. Let u satisfy (22), $u = 0$ on $\partial\Omega \backslash B_{R'}$, $u \in H^1_{\mathrm{loc}}(\Omega \backslash B_{R'})$. Then

$$|u(x)| \leqslant Ce^{-\alpha|x|}, \quad C, \alpha = const > 0. \qquad (28)$$

Proof. Let $N > R'$ and let v_N be the solution of the problem (24),(25). It is easy to prove as in Theorem 3, using $V(x) = const.$, that $|v_N(x)| \leqslant C_1$, where $C_1 = const$ and C_1 does not depend on N. Taking $\varphi = v_N e^{\alpha|x|}\Psi^2(x)$ as a test function for (24), where $\Psi = 1$ for $|x| > R' + 1$, $\Psi = 0$ for $|x| < R' + \frac{1}{2}$, $0 \leqslant \Psi \leqslant 1$, $\Psi \in C^\infty(\mathbf{R}^n)$, we obtain

$$C_1 \int\limits_{\Omega_N^{R'}} |\nabla v_N|^2 e^{\alpha|x|}\Psi^2(x)\mathrm{d}x + a_0 \int\limits_{\Omega_N^{R'}} |v_N|^{p+1}e^{\alpha|x|}\Psi^2(x)\mathrm{d}x$$

$$\leqslant \alpha C_2 \int\limits_{\Omega_N^{R'}} |v_N|^2 e^{\alpha|x|}\Psi^2\mathrm{d}x + \alpha \int\limits_{\Omega_N^{R'}} |\nabla v_N|^2 e^{\alpha|x|}\Psi^2\mathrm{d}x$$

$$+ 2 \int\limits_{\Omega_N^{R'+1/2}\backslash\Omega_N^{R'+1}} |\nabla v_N||v_N|e^{\alpha|x|}\Psi|\nabla\Psi|\mathrm{d}x. \qquad (29)$$

In order to estimate the first integral on the right-hand side of (29) we use the Friedrichs inequality

$$\int\limits_{\Omega_N^{R'}} |v_N|^2 e^{\alpha|x|}\mathrm{d}x \leqslant C_3 \int\limits_{\Omega_N^{R'}} |\nabla v_N|^2 e^{\alpha|x|}\Psi^2\mathrm{d}x, \qquad (30)$$

where as a consequence of (27) the constant C_3 is independent of N. We have

$$\int\limits_{\Omega_N^{R'+1/2}\backslash\Omega_N^{R'+1}} |\nabla v_N||v_N|e^{\alpha|x|}\Psi|\nabla\Psi|\mathrm{d}x$$

$$\leqslant \epsilon \int\limits_{\Omega_N^{R'+1/2} \backslash \Omega_N^{R'+1}} |\nabla v_N|^2 e^{\alpha|x|} |\Psi|^2 \mathrm{d}x$$

$$+ C_4 \int\limits_{\Omega_N^{R'+1/2} \backslash \Omega_N^{R'+1}} |v_N|^2 e^{\alpha|x|} |\nabla\Psi|^2 \mathrm{d}x$$

$$\leqslant \epsilon \int\limits_{\Omega_N^{R'+1/2} \backslash \Omega_N^{R'+1}} |\nabla v_N|^2 e^{\alpha|x|} |\Psi|^2 \mathrm{d}x$$

$$+ C_5(\epsilon) \int\limits_{\Omega_N^{R'+1/2} \backslash \Omega_N^{R'+1}} |v_N|^2 e^{\alpha|x|} \mathrm{d}x, \tag{31}$$

where ϵ is an arbitrarily small positive number.

For sufficiently small α we have from (28),(29),(31) that

$$\int\limits_{\Omega_N^{R'}} |\nabla v_N|^2 e^{\alpha|x|} \Psi^2 \mathrm{d}x + a_0 \int\limits_{\Omega_N^{R'}} |v_N|^{p+1} e^{\alpha|x|} \Psi^2 \mathrm{d}x$$

$$\leqslant C_6 \int\limits_{\Omega_N^{R'+1/2} \backslash \Omega_N^{R'+1}} |v_N|^2 e^{\alpha|x|} \mathrm{d}x. \tag{32}$$

The last integral in (32) is bounded by a constant which does not depend on N. In order to prove it we cover $\Omega_N^{R'+1/2} \backslash \Omega_N^{R'+1}$ by balls of radius $1/2$ and apply Theorem 1. From (32) we have

$$\int\limits_{\Omega_N^{R'+1}} |\nabla v_N|^2 e^{\alpha|x|} \mathrm{d}x + \int\limits_{\Omega_N^{R'+1}} |v_N|^{p+1} e^{\alpha|x|} \Psi^2 \mathrm{d}x \leqslant C_7, \tag{33}$$

where the constant C_7 does not depend on N and $\alpha = const > 0$ is sufficiently small.

Since $v_N = 0$ on $\partial\Omega$ for $|x| > R'$, by the De Giorgi type theorem (see e.g. [7], Theorem 8.17) we have

$$\sup_{x \in B_{1/2}(x_0) \cap \Omega_N^{R'}} |v_N(x)|$$

$$\leqslant C_8 \left[\int\limits_{B_1(x_0) \cap \Omega_N^{R'}} |v_N|^{p+1} \mathrm{d}x \right]^{\frac{1}{(1+p)}}$$

$$\leqslant C_9 \left[e^{-\alpha(|x_0|-1)} \int\limits_{B_1(x_0)\cap\Omega_N^{R'}} |v_N|^{p+1} e^{\alpha|x|} \mathrm{d}x \right]^{\frac{1}{(p+1)}}$$

$$\leqslant C_{10}\exp\{\alpha|x_0|\},$$

where $x_0 \in \Omega_N^{R'}$, $B_\rho(x_0) = \{x : |x - x_0| < \rho\}$, $\alpha = const > 0$ and C_{10} does not depend on N. Making $N \to \infty$ we obtain (28). This concludes the proof of the theorem.

The results of this section were obtained jointly with J.I. Diaz and published in a short communication [8].

References

[1] L. Veron, Comportement asymptotique des solutions d'équations elliptiques semi-lineaires dans $\mathbf{R}^n$, *Ann. Math. Pure. Appl.* (4), **127**, 1981, 25–50.

[2] H. Brezis, Semilinear equations in $\mathbf{R}^n$ without conditions at infinity, *Appl. Math. Optim.* **12**, 1984, 271–282.

[3] A. Brada, Comportement asymptotique des solutions d'équations elliptiques semi-lineaires dans un cylindre, *Thèse 3 cycle*, Univ. de Tours, 1987.

[4] V.A. Kondratiev , O.A. Oleinik, On asymptotic behavior of solutions of some nonlinear elliptic equations in unbounded domains, *Partial differential equations and related subjects, Proceddings of the Conference dedicated to Louis Nirenberg*, Longman, 1992, 163–195.

[5] H. Brezis, F. Browder, Strongly nonlinear elliptic boundary-value problems, *Ann. Scuola Norm. Sup. Pisa Cl. Sci.*, **5**, 1978, 587–603.

[6] J.I. Diaz, Nonlinear partial differential equations and free boundaries, *Res. Notes in math.*, **106**, Pitman, London, 1985.

[7] D. Gilbarg, N.S. Trudinger, Elliptic partial differential equations of second order, 2nd edition, Springer-Verlag, Berlin, 1983.

[8] J.I. Diaz, O.A. Oleinik, Nonlinear elliptic boundary-value problems in unbounded domains and the asymptotic behavior of its solutions, *C.R. Acad. Sci. Paris.*, **315** Serie 1, 1992, 787–792.

[9] B. Gidas, W.M. Ni, L. Nirenbrg, Symmetry and related properties via the maximum principle, *Comm. Math. Phys.* 68, 1979, 209–243.

[10] B. Gidas, W.M. Ni, L. Nirenberg, Symmetry of positive solutions of nonlinear elliptic equations in $\mathbf{R}^n$, *Math. Anal. and Applications, Part A, Advances in Math. Suppl. Studies 7A*, Academic Pr., 1981, 369–402.

[11] K. Kirchgassner, J. Scheurle, On the bounded solutions of a semilinear elliptic equation in a strip, *J. Diff. Equ.*, **32**, 1979, 119–148.

[12] V. Veron, Singular solutions of some nonlinear elliptic equations, *Nonlinear Anal. theory, Math. Appl.*, **5**, No 3, 1981, 225–242.

[13] P.L. Lions, Isolated singularities in semilinear problems, *J. Diff. Equations*, **38**, 1980, 441–450.

[14] C. Loewner, L. Nirenberg, Partial differential equations invariant under conformal or projective transformation, in "Contributions to Analysis", L. Alfors ed., 1974, 245–272.

[15] L. Veron, Singularités eliminable d'equaions elliptiques non linéaires, *J. Diff. Equations*, **41**, 9, 1981, 87–95.

[16] Y. Richard, L. Veron, Isotropic singularities of solutions of nonlinear elliptic inequalities, *Ann. Inst. Henri Poincaré*, **6**, 1989, 37–72.

[17] H. Brezis, L. Veron, Removable singularities for some nonlinear elliptic equations, *Arch. Rat. Mech. and Anal.*, **75**, No 1, 1980, 1–6.

[18] F. Bernis, Elliptic and parabolic semilinear problems without conditions at infinity, *Arch. Ration. Mech. and Anal.*, **106**, No 3, 1989, 217–241.

[19] J.B. Keller, On solutions of $\Delta u = f(u)$, *Comm. Pure and Appl. Math.*, **10**, No 4., 1957, 503–510.

[20] J.L. Vasques, L. Veron, Isolated singularities of some semilinear elliptic equations, *J. Diff. Equations*, **60**, 1985, 301–321.

[21] E.B. Dynkin. Superprocesses and partial differential equations, *Annals of probability*, **21**, 1993, 1185–1262.

[22] E.B. Dynkin. A probabilistic approach to one class of nonlinear differential equations, *Probab. Theory. Rel. Fields*, **89**, 1991, 89–115.

[23] V.A. Kondratiev, O.A. Oleinik, On asymptotic of solutions of nonlinear elliptic equations, *Uspechi Mat. Nauk*, **48**, No 4, 1993, 184–185.

[24] H. Berestycki, L. Nirenberg, Some qualitative properties of solutions of semilinear elliptic equations in cylindrical domains, *J. Geometry and Phys.*, ed. by P. Rabinowitz, Academic Press, 1990, 115–164.

[25] T. Kato, Schrödinder operators with singular potential, *Israel J. Math.*, **13**, 1972, 135–148.

[26] Ph. Benilan, H. Brezis, M. Crandall, A semilinear equation in $L^1(\mathbf{R}^N)$, *Ann. Sc. Norm. Sup. Pisa*, **2**, 1975, 523–555.

[27] Th. Gallouet, J.M. Morel, Resolution of a semilinear equation in L^1, *Proc. Royal Soc. Edinburgh*, **96A**, 1984, 275–288.

[28] H. Berestycki, L. Nirenberg, Asymptotic behavior via the harnacr inequality, in: nonlinear analysis, ed. by A. Ambrosetti, *Sc. Norm. Sup. Pisa*, 1991, 135–144

[29] H. Berestycki, L. Nirenberg, Travelling waves in cylinders, *Ann. Inst. H. Poicaré, Analyse non linéaire*, **9**, No 5, 1992, 497–572.

Chapter 2

On asymptotic behavior of solutions of some nonlinear elliptic equations in cylindrical domains

2.1 On the limits at infinity of solutions of some semi-linear second order elliptic equations in cylindrical domains.

Many problems of mathematical physics lead one to consider solutions of nonlinear partial differential equations in cylindrical and half-cylindrical domains and to study the behavior of solutions at infinity (stationary states, travelling waves, homogenization, boundary layer problem, Saint-Venant's principle and so on). In the paper by H. Berestycki and L. Nirenberg [1] monotonicity, symmetry, asymptotic properties for some solutions in cylindrical and half-cylindrical domains are proved for a class of nonlinear elliptic equations. We shall formulate here some of their results about asymptotic behavior of solutions at infinity.

Let us denote by $S(a, b)$ a domain $\{x : x \in \mathbf{R}^n, x' \in \omega, a < x_n < b\}$, where $x' = (x_1, \ldots, x_{n-1})$, ω is a bounded domain in $\mathbf{R}^{n-1} = (x_1, \ldots, x_{n-1})$ with a smooth boundary. In paper [1] the equation

$$\Delta u + g(x', u) = 0 \quad \text{in } S(0, +\infty), \qquad (1)$$

is considered (also some more general equations) with the boundary condition

$$\frac{\partial u}{\partial \nu} = 0 \quad \text{on } \sigma(0, +\infty), \qquad (2)$$

where ν is the exterior unit normal to $\sigma(0, +\infty)$, $\sigma(a, b) = \partial S(a, b) \cap \{x : a < x_n < b\}$, $\partial S(a, b)$ is the boundary of $S(a, b)$ (the Neumann problem) or with the boundary condition

$$u = 0 \text{ on } \sigma(0, +\infty) \tag{3}$$

(the Dirichlet problem).

They suppose that $u(x)$ is a solution of problem (1),(2) or (1),(3), $u \in C^2(S(0, +\infty))$, and is positive in $S(0, +\infty)$. Set

$$g_u(x', 0) = -a(x').$$

It is assumed that $g(x', u)$ is a Lipschitz continuous function in $\omega \times [0, S_0)$, where $S_0 = \sup u(x)$, $g(x', 0) = 0$, $g(x', u)$ is differentiable in $S(0, +\infty)$ at $u = 0$,

$$|g(x', u) + a(x')u| \leqslant M|u|^{1+\delta}, \quad M, \delta = const > 0.$$

Proposition 1. *Suppose that u is a positive solution of problem (1),(2) and $u(x', x_n) \to 0$ (uniformly for $x' \in \omega$) as $x_n \to \infty$ and the first eigenvalue μ_1 of the Neumann eigenvalue problem in ω*

$$(-\Delta + a(x'))v = \mu_1 v \text{ in } \omega, \tag{4}$$

$$\frac{\partial v}{\partial \nu} = 0 \text{ on } \partial\omega, \tag{5}$$

is positive.

Then there exists a positive constant α such that as $x_n \to \infty$

$$u(x', x_n) = \alpha \exp\{-\sqrt{\mu_1}x_n\}\Phi(x') + o(\exp\{-\sqrt{\mu_1}x_n\}), \tag{6}$$

$$\frac{\partial u}{\partial x_n} = -\alpha\sqrt{\mu_1} \exp\{-\sqrt{\mu_1}x_n\}\Phi(x') + o(\exp\{-\sqrt{\mu_1}x_n\}), \tag{7}$$

where $\Phi(x')$ is an eigenfunction of problem (4),(5), corresponding to the first eigenvalue μ_1.

H. Berestycki and L. Nirenberg note that the equation

$$\Delta u - 2u^3 = 0$$

has a solution $u(x', x_n) = (1 + x_n)^{-1}$ with the boundary condition (2).

Proposition 2. *Let $u(x', x_n)$ be a solution of problem (1),(3) in $S(0, +\infty)$, $u(x)$ be positive in $S(0, +\infty)$, $u(x', x_n) \to 0$ as $x_n \to \infty$, $|g_u(x', u) - g_u(x', 0)| \leqslant M|u|^\delta$ for $0 \leqslant u \leqslant S_0$, $\delta = const > 0$, $S_0 = \sup_{S(0,\infty)} u$. Then* *there exists a constant α such that (6) and (7) are valid with $\mu_1 = \lambda_1$, where λ_1 and $\Phi(x')$ are the first eigenvalue and corresponding eigenfunction of the eigenvalue problem*

$$(-\Delta + a(x'))w = \lambda_1 w \quad in \ \omega,$$

$$w = 0 \quad on \ \partial\omega,$$

if λ_1 is positive.

The following questions arise:

(1) What can happen, when $\mu_1 = 0$ or $\lambda_1 = 0$?

(2) Are theorems of the Phragmen–Lindelöf type valid for the solutions of problem (1),(2) or (1),(3)?

(3) What is the asymptotic behavior at infinity of solutions of the equation (1), with boundary conditions (2) or (3), which change sign?

We shall consider some of these problems in this lecture.

Lemma 1. *Suppose that $u \in C^2(S(-\infty,\infty))$, $u = 0$ on $\partial S(-\infty,+\infty)$, $u \in C(\overline{S}(-\infty,+\infty))$ or $\dfrac{\partial u}{\partial \mu} = 0$ on $\sigma(-\infty,+\infty)$ and $u \in C^1(\overline{S}(-\infty,+\infty))$,*

$$\sum_{i,j=1}^{n} a_{ij}(x)\frac{\partial^2 u}{\partial x_i \partial x_j} - f(x,u) \geqslant -A|\nabla u|^{p_1} - B|\nabla u|, \qquad (8)$$

where

$$\frac{\partial u}{\partial \mu} \equiv \sum_{i,j}^{n} a_{ij}\frac{\partial u}{\partial x_i}\nu_j = 0,$$

$$\alpha_1|\xi|^2 \leqslant \sum_{i,j=1}^{n} a_{ij}(x)\xi_i\xi_j \leqslant \alpha_2|\xi|^2, \alpha_1,\alpha_2 = const > 0, \forall \xi \in \mathbf{R}^n,$$

and the following further assumptions are made: a_{ij}, b_i are bounded measurable functions, $f(x,u_1) - f(x,u_2) > 0$ for $u_1 > u_2 \geqslant 0$, $f(x,u) \geqslant f_1(u)$, $f_1(y) \geqslant a_1 y^p$ for $y \geqslant 0$, $a_1 = const > 0$, $A, B = const \geqslant 0$, $p > 1$, $p_1 = 2p/(1+p)$ and $f_1(y)$ is a continuous function. Then $u(x)$ cannot attain positive values in $S(-\infty,\infty)$.

In order to prove Lemma 1 we need to prove the following auxiliary result.

Lemma 2. *Let* $y(t)$ *be a solution of the equation*

$$y'' + a_1|y'|^{p_1} + a_2|y'| = f(y), \qquad a_1, a_2 = const \geqslant 0 , \qquad (9)$$

with initial conditions

$$y(0) = \alpha > 0, y'(0) = 0, t \geqslant 0 ,$$

$f(y) > 0$ *for* $y > 0$, $f(y) \geqslant a_3 y^p$, $p > 1$, $a_3 = const > 0$, *for* $y > 1$, $p_1 = \dfrac{2p}{1+p}$, $f(y)$ *is a continuous function for all* y.

    Then there exists $T(\alpha) > 0$ *such that* $y(t) \to \infty$ *as* $t \to T(\alpha) - 0$, $y(t) > 0$ *for* $(0, T)$ *and* $y'(t) > 0$ *for* $(0, T)$.

Proof. It is easy to see that $y'(t) > 0$ for $t > 0$, $y(t) > \alpha$ for $t > 0$. Indeed, we have $y'(t) > 0$ for small t, since $f(\alpha) > 0$. If $y'(t^0) = 0$ for $t^0 > 0$, then the equation (9) cannot be satisfied at $t = t^0$. Let us change variables in (9):

$$y' = u(y) .$$

Then we have

$$y'' = u'u, \qquad uu' + a_1|u|^{p_1} + a_2|u| = f(y), y > \alpha, u(\alpha) = 0 .$$

Set $u^2 = z$. Then

$$\frac{1}{2}z' + a_1|z|^{\frac{p_1}{2}} + a_2|z|^{\frac{1}{2}} = f(y), y > \alpha, z(\alpha) = 0 . \qquad (10)$$

For $w = \beta^{-1}z$ we have from (10),

$$\frac{1}{2}w' + a_1\beta^{\frac{p_1}{2}-1}|w|^{\frac{p_1}{2}} + a_2\beta^{\frac{1}{2}-1}|w|^{\frac{1}{2}} = \beta^{-1}f(y) . \qquad (11)$$

Since $|w|^{\frac{1}{2}} \leqslant 1 + |w|^{\frac{p_1}{2}}$, it follows from (11) that

$$\frac{1}{2}w' + a_4|w|^{\frac{p_1}{2}} + \left(a_1\beta^{\frac{p_1}{2}-1} - a_4\right)|w|^{\frac{p_1}{2}} + a_2\beta^{\frac{1}{2}-1}\left(1 + |w|^{\frac{p_1}{2}}\right)$$
$$\geqslant \beta^{-1}f(y) \geqslant \beta^{-1}(k + a_5 y^p) ,$$

where k and a_4, a_5 are positive constants, k and a_5 exist because of the properties of function $f(y)$ for $y \geqslant \alpha > 0$. Let us take $\beta > 0$ so small that $a_2\beta^{\frac{1}{2}} < k$ and a_4 so large that $a_1\beta^{\frac{p_1}{2}-1} < a_4$. Then we have

$$\frac{1}{2}w' + a_6|w|^{\frac{p_1}{2}} \geqslant a_7 y^p , a_7 = \beta^{-1}a_5 , a_6 = a_4 + a_2\beta^{\frac{1}{2}-1} . \qquad (12)$$

Now we introduce new variables

$$w = vy^{p+1} \ , y = e^\tau \ .$$

We get from (12) that

$$\frac{1}{2}yv'_y + \frac{1}{2}(p+1)v + a_6|v|^{\frac{p_1}{2}} \geqslant a_7 \ ,$$

$$\frac{1}{2}v'_\tau + \frac{1}{2}(p+1)v + a_6|v|^{\frac{p_1}{2}} \geqslant a_7 \ ,$$

$$v(\ln \alpha) = 0 \ .$$

Since v'_τ is positive for $|v|$ sufficiently small, we have $v \geqslant \delta = const > 0$ for $\tau \geqslant \tau_0 > 0$. It means that $w \geqslant \delta y^{p+1}$, $z \geqslant \beta\delta y^{p+1}$, $u \geqslant \beta^{\frac{1}{2}}\delta^{\frac{1}{2}}y^{(p+1)/2}$ and $y' \geqslant \delta^{\frac{1}{2}}\beta^{\frac{1}{2}}y^{(p+1)/2}$ for $t \geqslant t_0$. Integrating the last inequality for y from t_0 to t, we obtain

$$\left(\frac{-p+1}{2}\right)^{-1} \left(y(t)^{-\frac{p}{2}+\frac{1}{2}} - y(t_0)^{-\frac{p}{2}+\frac{1}{2}}\right) \geqslant (t-t_0)\delta^{\frac{1}{2}}\beta^{\frac{1}{2}} \ , y(t_0) = e^{\tau_0} \ . \quad (13)$$

It follows from (13) that $y(t) \to \infty$ as $t \to T - 0$. The lemma is proved.

Proof of Lemma 1. Suppose that $u(x'_0, 0) = 2\alpha > 0$ for some $(x'_0, 0) \in S(-\infty, +\infty)$. Let us take the solution $y(x_n)$, constructed in Lemma 2 such that $y(0) = \alpha$, $y'(0) = 0$, and the equation

$$y'' + a(|y'|^{p_1} + |y'|) = bf_1(y) \ ,$$

holds where $a, b = const > 0$. We have

$$y'' - \frac{1}{a_{nn}(x)}f(x, y) \quad \leqslant \quad y'' - bf_1(y) = -a|y'|^{p_1} - a|y'|$$

$$\leqslant \quad -\frac{A}{a_{nn}(x)}|y'|^{p_1} - \frac{B}{a_{nn}(x)}|y'| \ ,$$

if a^{-1}, b are sufficiently small. Therefore,

$$a_{nn}(x)y'' - f(x, y) \leqslant -A|y'|^{p_1} - B|y'| \ . \quad (14)$$

Set $w = y - u$. Then from (8) and (14) we have

$$\sum_{i,j=1}^{n} a_{ij}(x)\frac{\partial^2 w}{\partial x_i \partial x_j} - (f(x, y) - f(x, u))$$

$$\leqslant -A(|y'|^{p_1} - |\nabla u|^{p_1}) - B(|y'| - |\nabla u|) \ . \quad (15)$$

Consider the inequality (15) in $S(-T,T)$ and define $y(x_n)$ for $-T < x_n < 0$ as $y(x_n) = y(-x_n)$. We have $w(x',x_n) \to \infty$ as $x_n \to T - 0$ and $x_n \to -T + 0$, $w(x'_0, 0) = -\alpha < 0$. Since we assume $u(x'_0, 0) = 2\alpha$, there exists a point $\bar{x}$, where w attains the least negative value. If $\bar{x} \in S(-T,T)$, then $\dfrac{\partial w}{\partial x_i} = 0$ at $\bar{x}$, $\dfrac{\partial y}{\partial x_i} = \dfrac{\partial u}{\partial x_i}$ at $\bar{x}$, $i = 1,\ldots,n$, $y(\bar{x}_n) < u(\bar{x})$. This contradicts (15). Indeed, from (15) we have

$$\sum_{i,j=1}^{n} a_{ij}(x)\frac{\partial^2 w}{\partial x_i \partial x_j} - \frac{f(x,y) - f(x,u)}{y - u} w \leqslant 0 \qquad (16)$$

at $\bar{x}$. Since $f(x,y)$ is a monotonic function of y, the left hand side of (15) must be positive at $\bar{x}$.

In the case of the Dirichlet condition $u = 0$ on $\partial S(-\infty, +\infty)$ the least negative value of w cannot be attained on $\sigma(-T, +T)$, where $w = y > 0$, and at $x_n = \pm T$, where $w = +\infty$. In the case of the Neumann condition $\dfrac{\partial u}{\partial \mu} = 0$ on $\partial S(-\infty, +\infty)$, if the least negative value of w is attained on $\sigma(-\infty, +\infty)$, then $\frac{\partial w(\bar{x})}{\partial \mu} = 0$, since $\frac{\partial y(\bar{x})}{\partial \mu} = 0$, $\frac{\partial u(\bar{x})}{\partial \mu} = 0$, and $|\nabla u(\bar{x})| = |\nabla y(\bar{x})|$. Therefore, the right hand side of (15) is small in a neighbourhood of $\bar{x}$ and we can apply the lemma on the normal derivative ([2],[3]) to the equation

$$\sum_{i,j=1}^{n} a_{ij}(x)\frac{\partial^2 w}{\partial x_i \partial x_j} - \frac{f(x,y) - f(x,u)}{2(y - u)} w$$
$$\equiv \Phi(x)$$
$$\leqslant \frac{1}{2}\frac{f(x,y) - f(x,u)}{(y - u)} w - A(|y'|^p - |\nabla u|^p) - B(|y'| - |\nabla u|)$$
$$< 0$$

in the neighborhood of $\bar{x}$. According to this lemma $\dfrac{\partial w(\bar{x})}{\partial \nu} < 0$. These contradictions to the assumption $u(x'_0, 0) = 2\alpha > 0$ prove Lemma 1.

Corollary 1. *Suppose that in Lemma 1 instead of (8) we have*

$$\sum_{i,j=1}^{n} a_{ij}(x)\frac{\partial^2 u}{\partial x_i \partial x_j} - f(x,u) \leqslant A|\nabla u|^{p_1} + B|\nabla u|, \qquad (17)$$

and that $-f(x, -u)$, a_{ij}, p_1, p, A, B *satisfy the conditions of Lemma 1. Then* $u(x)$ *cannot attain negative values in* $S(-\infty, +\infty)$.

This is evident since the function $-u$ satisfies the conditions of Lemma 1.

Theorem 1. *Suppose that $u \in C^2(S(-\infty, +\infty))$, $u = 0$ on $\partial S(-\infty, +\infty)$ and $u \in C(\overline{S}(-\infty, +\infty))$ or $\dfrac{\partial u}{\partial \mu} = 0$ on $\partial S(-\infty, +\infty)$ and $u \in C^1(\overline{S}(-\infty, +\infty))$,*

$$\left| \sum_{i,j=1}^n a_{ij}(x) \frac{\partial^2 u}{\partial x_i \partial x_j} - f(x, u) \right| \leqslant A|\nabla u|^{p_1} + B|\nabla u| \,, \qquad (18)$$

where a_{ij}, p_1, A, B, p, $f(x, u)$, $-f(x, -u)$ satisfy the conditions of Lemma 1. Then $u \equiv 0$ in $S(-\infty, +\infty)$.

Theorem 2. *If $u(x)$ satisfies the inequality (18) in $S(0, \infty)$ with the boundary condition $\frac{\partial u}{\partial \mu} = 0$ or $u = 0$ on $\sigma(0, \infty)$, then $u(x', x_n) \to 0$ as $x_n \to \infty$.*

Proof. Assume that there exists a sequence $(x'(k), x_n^k)$, $k \to \infty$, such that $u(x'(k), x_n^k) > 2\alpha > 0$ as $x_n^k \to \infty$, $\alpha = const > 0$. Consider the functions $y(x_n - x_n^k)$, constructed in Lemma 2. Then $y(x_n - x_n^k) \to \infty$ as $x_n - x_n^k \to T$ and $x_n - x_n^k \to -T$. Suppose that $x_n^k - T > 0$ and consider the function $w = y - u$ in the domain $S(x_n^k - T, x_n^k + T)$. It is easy to see that $w \to \infty$ as $x_n^k \to \pm T$ and $w(x'(k), x_n^k) < 0$. If we consider the inequality (15) for w, in $S(x_n^k - T, x_n^k + T)$, we get a contradiction in the same way as in the proof of Lemma 1. Next, if we assume that $u(x'(k), x_n^k) < -2\alpha$ for some sequence x_n^k, then we get a contradiction by considering the function $-u$. This completes the proof.

The next result gives an estimate for the decay of solutions, considered in Theorem 2.

Theorem 3. *Let $u(x)$ satisfy the inequality*

$$\left| \sum_{i,j=1}^n a_{ij}(x) u_{x_i x_j} - f(x, u) \right| \leqslant A|\nabla u|^{p_1} + B|\nabla_{x'} u| \qquad (19)$$

in $S(0, \infty)$ and condition $\frac{\partial u}{\partial \mu} = 0$ on $\sigma(0, \infty)$, where $A, B = const \geqslant 0$, $\nabla_{x'} u = (u_{x_1}, \cdots, u_{x_{n-1}})$, $p_1 = \frac{2p}{p+1}$, $p > 1$, $a_{ij}(x)$, $f(x, u)$ and $-f(x, -u)$ satisfy the conditions of Lemma 1. Then

$$|u(x)| \leqslant C|x_n|^{\frac{2}{1-p}} \,, \qquad (20)$$

where $C = const > 0$.

Proof. Consider the function $z(x_n) = C_1 x_n^\gamma$, $\gamma = \frac{2}{1-p}$. It is easy to see that for positive constants a and b,

$$z'' + a|z'|^{p_1} - bz^p = C_1\gamma(\gamma - 1)x_n^{\gamma-2} + a|\gamma C_1 x_n^{\gamma-1}|^{p_1} - bC_1^p x_n^{\gamma p} \leqslant 0 ,$$

if C_1 is sufficiently large, since $p > p_1$ for $p > 1$. We have

$$z'' - \frac{1}{a_{nn}}f(x,z) \leqslant z'' - bz^p \leqslant -a|z'|^{p_1} \leqslant -\frac{A}{a_{nn}}|z'|^{p_1} , \qquad (21)$$

if b and a^{-1} are sufficiently small. For the function $w = z - u$ we get

$$\sum_{i,j=1}^{n} a_{ij}(x)\frac{\partial^2 w}{\partial x_i \partial x_j} - \frac{f(x,z) - f(x,u)}{z - u}w$$
$$\leqslant -A(|z'|^{p_1} - |\nabla u|^{p_1}) - B(|\nabla_{x'} u|) \qquad (22)$$

in $S(0,\infty)$. We prove that $w \geqslant 0$. If w takes negative values, then there exists a point $\overline{x}$ in $\overline{S}(0, M)$, $M = const > 0$, where it attains the least negative value. The point $\overline{x}$ cannot belong to $S(0, M)$, since that would contradict (22), and $w \to \infty$ as $x_n \to 0$. If $\overline{x} \in \sigma(0, M)$, then in the case of the Neumann problem we have a contradiction in the same way as in the proof of Lemma 1. For the case of the Dirichlet problem we have $w > 0$ on $\sigma(0,M)$. Let us take M so large that $|w| < \epsilon$ for $x_n = M$. Then we have $w \geqslant -\epsilon$ in $S(0, M)$. Since ϵ is an arbitrary number, $w \geqslant 0$ in $S(0, M)$. This means that $u \leqslant C_1 x_n^\gamma$ in $S(0, M)$. In a similar way, considering the function $u_1 = -u$, we get that $u + z \geqslant 0$ and $-z \geqslant u$ in $S(0,\infty)$. The theorem is proved.

It is easy to see that the equation

$$\Delta u - k|u|^{p-1}u = 0 \text{ in } S(0, +\infty) , \qquad (23)$$

where $k = const > 0$, has a solution

$$u(x) = C_0 x_n^{\frac{2}{1-p}} , \quad \frac{\partial u}{\partial \mu} = 0 \text{ on } \sigma(0, \infty),$$

where

$$C_0 = \left(\frac{2(1+p)}{(p-1)^2 k}\right)^{\frac{1}{p-1}}$$

This means that the estimate (20) cannot be improved for all classes of inequalities (19).

We set $\omega(\tau) = S(-\infty, +\infty) \cap \{x : x_n = \tau\}$.

Theorem 4. *Let $u(x)$ be a solution of the equation*

$$\sum_{i,j=1}^{n} \frac{\partial}{\partial x_i}\left(a_{ij}(x)\frac{\partial u}{\partial x_j}\right) - f(x,u) = 0 \tag{24}$$

in $S(0,\infty)$, $f(x,u)u \geqslant 0$ and $u = 0$ on $\sigma(0,\infty)$.
For some positive constants λ_1, λ_2 and $\forall \xi \in \mathbf{R}^n$,

$$\lambda_1|\xi|^2 \leqslant \sum_{i,j=1}^{n} a_{ij}(x)\xi_i\xi_j \leqslant \lambda_2|\xi|^2, \qquad \lambda_1,\lambda_2 = const > 0, \tag{25}$$

and $f(x,u)$ satisfies either the conditions of Theorem 1 or $|u(x)| \leqslant C|x|^\beta$ for $|x| > a$, where C,β,a are positive constants and $\beta < 1/2$. Then

$$\int_{\omega(\tau)} |u(x)|^2 dx' \leqslant C_1 \exp(-\alpha\tau), \qquad C_1,\alpha = const > 0.$$

Proof. Let $\phi(t)$ be a C^∞-function such that $\phi(t) = 1$, for $|t| < 1$, $\phi(t) = 0$ for $|t| > 2$, and $0 \leqslant \phi \leqslant 1$. Multiplying equation (24) by $u(x)\phi^2(\frac{x_n}{N})$, $N = const > 1$, integrating the equation obtained over $S(\tau,\infty)$, and transforming the first integral by integration by parts, we get, using (25),

$$\int_{S(\tau,\infty)} \lambda_1|\nabla u|^2\phi^2 dx + \int_{S(\tau,\infty)} f(x,u)u\phi^2 dx$$

$$\leqslant \left|\int_{S(\tau,\infty)} \sum_{i=1}^{n} \frac{\partial u}{\partial x_i}u \cdot 2\phi\frac{\partial\phi}{\partial x_i}dx\right| + \left|\int_{\omega(\tau)} \frac{\partial u}{\partial\gamma}u\phi^2 dx'\right|. \tag{26}$$

It is easy to see that

$$\left|\int_{S(\tau,\infty)} 2\sum_{i=1}^{n} \frac{\partial u}{\partial x_i}u\phi\frac{\partial\phi}{\partial x_i}dx\right|$$

$$\leqslant \epsilon \int_{S(\tau,\infty)} \sum_{i=1}^{n} \left|\frac{\partial u}{\partial x_i}\right|^2\phi^2 dx + \frac{1}{\epsilon}\int_{S(N,2N)} |u|^2(\phi')^2 N^{-2}dx,$$

$$\epsilon = const > 0, \tag{27}$$

and the last integral tends to zero as $N \to \infty$ on account of Theorem 2 or the assumption $|u(x)| < C|x|^{\beta}$. From (26) as $N \to \infty$, it follows that

$$
\int\limits_{S(\tau,\infty)} |\nabla u|^2 dx \leqslant C_2 \left(\int\limits_{\omega(\tau)} |\nabla u|^2 dx' \right)^{\frac{1}{2}} \left(\int\limits_{\omega(\tau)} |u|^2 dx' \right)^{\frac{1}{2}}. \tag{28}
$$

Since $u = 0$ on $\partial\omega(\tau)$, according to the Friedrichs inequality we have

$$
\int\limits_{\omega(\tau)} u^2 dx' \leqslant C_3(\omega(\tau)) \int\limits_{\omega(\tau)} |\nabla u|^2 dx', \qquad C(\omega(\tau)) = const > 0. \tag{29}
$$

From (28) and (29) it follows that

$$
\int\limits_{S(\tau,\infty)} |\nabla u|^2 dx \leqslant C_4 \int\limits_{\omega(\tau)} |\nabla u|^2 dx', \qquad C_t = const > 0. \tag{30}
$$

Setting

$$
J(\tau) \equiv \int\limits_{S(\tau,\infty)} |\nabla u|^2 dx,
$$

the inequality (30) can be written in the form

$$
J \leqslant -C_4 \frac{\partial J}{\partial \tau}. \tag{31}
$$

Integrating this inequality, we obtain

$$
\int\limits_{S(\tau,\infty)} |\nabla u|^2 dx \leqslant C_5 \exp\{-C_6\tau\}, \tag{32}
$$

where the positive constants C_5, C_6 do not depend on τ. By the mean value theorem and (29) we have

$$
\int\limits_{\omega(\tau')} |u|^2 dx' \leqslant C_3 \int\limits_{\omega(\tau')} |\nabla u|^2 dx'
$$

$$
\leqslant C_3 \int\limits_{S(\tau,\tau+1)} |\nabla u|^2 dx \leqslant C_\tau \exp\{-C_6\tau\} \tag{33}
$$

for some τ', $\tau \leqslant \tau' \leqslant \tau + 1$. For any τ,

$$
\begin{aligned}
\int_{\omega(\tau)} |u|^2 \mathrm{d}x' &= \int_{\omega(\tau')} |u|^2 \mathrm{d}x' - \int_{S(\tau,\tau')} 2u \frac{\partial u}{\partial x_n} \mathrm{d}x \\
&\leqslant \int_{\omega(\tau)} |u|^2 \mathrm{d}x' \\
&\quad + 2 \left(\int_{S(\tau,\tau')} |u|^2 \mathrm{d}x \right)^{\frac{1}{2}} \left(\int_{S(\tau,\tau')} \left(\frac{\partial u}{\partial x_n} \right)^2 \mathrm{d}x \right)^{\frac{1}{2}} \\
&\leqslant C_8 \exp\{-C_6 \tau\}.
\end{aligned}
$$

This concludes the proof of the theorem.

Theorem 5. *Let $u(x)$ be a solution of equation (24) in $S(0,\infty)$, $u = 0$ on $\sigma(0,\infty)$, $f(x,u)$ satisfies the conditions of Theorem 4, $|f(x,u)| \leqslant C_1|u|^r$, $r = const > 0$, $\tau \leqslant \frac{4}{n}$. Then*

$$
|u(x)| \leqslant C_1 \exp\{-\alpha x_n\}, \qquad C_1, \alpha = const > 0. \tag{34}
$$

Proof. As in the proof of Theorem 4 we obtain the inequality (32). From (29) and (32) we have

$$
\int_{S(\tau,\tau+1)} |u|^2 \mathrm{d}x \leqslant C_5 \int_{S(\tau,\tau+1)} |\nabla u|^2 \mathrm{d}x \leqslant C_8 \exp\{-C_6 \tau\}. \tag{35}
$$

From the De Giorgi type theorem (see, for example, [4],[5]) we get

$$
\max |u(x)|^2_{S(\tau+1/3,\tau+2/3)}
$$
$$
\leqslant C_9 \left(\int_{S(\tau,\tau+1)} |u|^2 \mathrm{d}x + \left(\int_{S(\tau,\tau+1)} |f(x,u)|^q \mathrm{d}x \right)^{2/q} \right),
$$
$$
q \geqslant \frac{n}{2}. \tag{36}
$$

It is easy to see that

$$
\int_{S(\tau,\tau+1)} |f(x,u)|^q \mathrm{d}x
$$

$$\leqslant C_1^q \int\limits_{S(\tau,\tau+1)} |u|^{rq}\mathrm{d}x$$

$$\leqslant C_{10} \int\limits_{S(\tau,\tau+1)} |u|^2\mathrm{d}x$$

$$\leqslant C_{11} \exp\{-C_6\tau\} \tag{37}$$

since we can suppose that $rq \leqslant 2$. From (35)–(37) we get (34).

Now we consider solutions of equation (24) in $S(0,\infty)$ with the boundary condition $\frac{\partial u}{\partial \mu} = 0$ on $\sigma(0,\infty)$.

We say that a solution $u(x)$ of equation (24) changes sign in $S(0,\infty)$ if $u(x)$ attains positive and negative values in $S(k,\infty)$ for any $k = const > 0$. From the maximum principle it follows that $u = 0$ at (x_0', x_n) for any $x_n > 0$ and some x_0', which depends on x_n, since if $u > 0$ in $\omega(\tau)$ and $u \to 0$ as $x_n \to \infty$, then $u > 0$ in $S(\tau,\infty)$.

Theorem 6. *Let $u(x)$ be a solution of equation (24) with the boundary condition*

$$\frac{\partial u}{\partial \mu} = 0 \ \ on \ \sigma(0,\infty),$$

$f(x,u)u \geqslant 0$ and $u(x) \to 0$ as $x_n \to \infty$, $n = 2$, $u(x)$ change sign in $S(k,\infty)$ for any $k > 0$.

Then there exist constants C and $\beta > 0$ such that

$$|u(x_1,x_2)| \leqslant C\exp\{-\beta x_2\}. \tag{38}$$

Proof. In the case $n = 2$ the proof of Theorem 6 is similar to the proof of Theorem 4. The inequality (29) can be obtained using the fact that for any $\tau > 0$ there exists x_0' such that $u(x_0',\tau) = 0$ and

$$|u(x_1,\tau)|^2 = \left| \int\limits_{x_0'}^{x_1} \frac{\partial u(x_1,\tau)}{\partial x_1}\mathrm{d}x_1 \right|^2$$

$$\leqslant C_1 \int\limits_{x_0'}^{x_1} |\nabla u|^2\mathrm{d}x_1$$

$$\leqslant C_1 \int\limits_{\omega(\tau)} |\nabla u|^2\mathrm{d}x_1,$$

$$\int_{\omega(\tau)} |u|^2 dx_1 \;\leqslant\; C_2 \int_{\omega(\tau)} |\nabla u|^2 dx_1, \qquad C_1, C_2 = const > 0.$$

Then in the same way as in the proof of Theorem 4, we get (33). It follows from (33) that

$$|u(x_1, \tau')| \leqslant \int_{\omega(\tau')} \left| \frac{\partial u}{\partial x_1} \right| dx_1 \leqslant C_3 \left(\int_{\omega(\tau')} |\nabla u|^2 dx_1 \right)^{\frac{1}{2}} \leqslant C_4 \exp\{-C_5 \tau'\},$$

(39)

where the constants C_3, C_4 do not depend on τ and τ'. Consider the intervals $(\tau - 2, \tau - 1)$ and $(\tau + 2, \tau + 3)$. We can find values τ' and τ'' such that $\tau - 2 < \tau' < \tau - 1$, $\tau + 2 < \tau'' < \tau + 3$ and for which (39) is valid. Then using the maximum principle we obtain the estimate (38) for $u(x_1, \tau)$. The theorem is proved.

For simplicity we consider now the model equation

$$\Delta u - Q(x)|u|^{p-1} u = 0,$$
$$Q(x) \geqslant C_0 = const > 0, \;\; Q(x) \leqslant C_1, \qquad (40)$$

in $S(0, \infty)$ with the boundary condition

$$\frac{\partial u}{\partial \nu} = 0 \;\; \text{on } \sigma(0, \infty). \qquad (41)$$

We study the behavior of solutions $u(x)$ of (40),(41) as $x_n \to +\infty$ in the case $n \geqslant 3$ and $u(x)$ changes sign in $S(k, \infty)$ for any $k > 0$. First we prove some lemmas.

Lemma 3. *Assume that $u(x)$ satisfies the equation*

$$\Delta u - Q(x)|u|^{p-1} u = F(x) \qquad (42)$$

in $S(0, \infty)$ with the boundary condition (41), $Q(x) \geqslant 0$ and

$$|u(x', x_n)| \;\leqslant\; C_1(1 + |x_n|)^\gamma,$$
$$|F(x)| \;\leqslant\; C_2(1 + |x_n|)^{\gamma - 2}, \;\; \gamma < \frac{1}{2}, C_1, C_2 = const > 0. \qquad (43)$$

Then

$$\int_{S(\tau, \infty)} |\nabla u|^2 dx \leqslant C_3 \tau^{2\gamma - 1} \qquad (44)$$

for some constant C_3.

Proof. Consider the function $\Theta(x_n)$ such that

$$\Theta(x_n) = \frac{2x_n - \tau}{\tau} \text{ for } x_n \in \left[\frac{\tau}{2}, \tau\right],$$

$$\Theta(x_n) = 1 \text{ for } x_n \in [\tau, \tau_1],$$

$$\Theta(x_n) = 0 \text{ for } x_n < \tau/2,$$

and for $x_n > 2\tau_1$, $\Theta = \dfrac{2\tau_1 - x_n}{\tau_1}$ for $x_n \in [\tau_1, 2\tau_1]$. Multiplying equation
(42) by $\Theta^2 u$, integrating it over $S(\frac{\tau}{2}, \infty)$ and transforming the first term
by integration by parts, we obtain

$$\int\limits_{S(\frac{\tau}{2},\infty)} \Theta^2 |\nabla u|^2 dx$$

$$\leqslant C \left(\int\limits_{S(\tau_1, 2\tau_1)} (\Theta')^2 |u|^2 dx + \int\limits_{S(\frac{\tau}{2},\infty)} \Theta^2 |u||F| dx \right.$$

$$\left. + \int\limits_{S(\frac{\tau}{2},\tau)} (\Theta')^2 |u|^2 dx \right).$$

Using (43), taking into account that $\Theta'(x_n) = -\frac{1}{\tau_1}$ for $x_n \in [\tau_1, 2\tau_1]$
and letting $\tau_1 \to \infty$, we get (44).

Lemma 4. *Let $u(x)$ be a solution of the equation*

$$\Delta u = F(x) \tag{45}$$

*in $S(0, \infty)$ with the boundary condition (41) and satisfying the following
conditions:*

$$|u(x)| \leqslant C_1 (1 + x_n)^\gamma, \quad \gamma < 1/2, \quad |F(x)| \leqslant C_2 (1 + x_n)^{\gamma - 2},$$

$$\int\limits_{S(\tau, \infty)} |\nabla u|^2 dx \leqslant C_3 \tau^{2\gamma - 1}, \quad C_1, C_2, C_3 = const > 0.$$

Then in $S(0, \infty)$

$$|\nabla u(x)| \leqslant C_4 (1 + x_n)^{\gamma - \frac{1}{2}}, \quad C_4 = const > 0. \tag{46}$$

Proof. For the solution of equation (45) with the boundary condition (41) we get, using the Green function, that

$$\max_{S(\tau,\tau+1)} |\nabla u(x)|^2$$

$$\leqslant C_5 \left[\max_{S(\frac{\tau}{2},\tau+2)} |F|^2 + \int\limits_{S(\frac{\tau}{2},\tau+2)} |u|^2 dx \right], \quad C_5 = const > 0. \quad (47)$$

In order to estimate the last integral we use the Poincaré inequality, since we can assume that

$$\int\limits_{S(\frac{\tau}{2},\tau+2)} u\, dx = 0.$$

The inequality (46) follows from (47) and the assumptions of the lemma.

Theorem 7. *Let $u(x)$ be a solution of the equation (40) in $S(0,\infty)$ with the boundary condition (41), $p > 1$. We assume that $u(x)$ changes sign in $S(k,\infty)$ for any $k > 0$. Then*

$$|u(x',x_n)| \leqslant C_h x_n^{\frac{2}{1-p}-h}, \quad C_h = const > 0, \quad (48)$$

where h is any positive number.

Proof. According to Theorem 3 we have

$$|u(x',x_n)| \leqslant C_1 x_n^{\frac{2}{1-p}}$$

and the conditions of Lemmas 3 and 4 are satisfied with $\gamma = \frac{2}{1-p}$. Therefore

$$|\nabla u(x)| \leqslant C_2(1+x_n)^{\gamma-\frac{1}{2}}, \quad (49)$$

where $\gamma = \frac{2}{1-p}$, $C_j = const > 0$.

Since we suppose that $u(x)$ changes sign, it means that for any $x_n > 0$ there exists a point (x'_0, x_n) such that $u(x'_0, x_n) = 0$. Then for any (x', x_n) we have

$$u(x',x_n) = \int\limits_{x'_0}^{x'} \frac{\partial u}{\partial l} dl, \quad (50)$$

where the integration is over the interval (x_0', x') on $\omega(x_n)$. From (50) and (49) we get (48) for $h = \frac{1}{2}$. Taking $\gamma = \frac{2}{1-p} - \frac{1}{2}$, using Lemmas 3, 4 and (50), we get the estimate (48) with $h = 1$. Using this argument repeatedly we get (48) for any $h > 0$.

The main theorem of this lecture is the following proposition.

Theorem 8. *Let $u(x)$ be a solution of the equation (40) in $S(0,\infty)$ with the boundary condition (41), $p > 1$. Suppose that $u(x)$ changes sign in $S(k,\infty)$ for any $k > 0$. Then*

$$|u(x', x_n)| \leqslant C \exp\{-\alpha x_n\},$$

$\alpha = const > 0$, $C = const > 0$.

Proof. From Theorem 7 it follows that $|u(x)| \leqslant C_1 |1 + x_n|^{-(2+\epsilon)}$ in $S(0,\infty)$, where $\epsilon = const > 0$. Consider the function $v = \Theta u$, where $\Theta(x_n) = 0$ for $x_n < T$, $T > 0$, $\Theta(x_n) = 1$ for $x_n > T + 1$, $\Theta(x_n) \in C^\infty(\mathbf{R}^1)$. For the function v we get the equation

$$\Delta v - g(x)v = F(x), \quad \text{in } S(-\infty, +\infty), \tag{51}$$

where $g(x) = 0$ for $x_n < T$ and $g(x) = |u|^{p-1} Q$ for $x_n \geqslant T + 1$ and $F(x) = \Delta\Theta u + 2\sum_{i=1}^{n} \dfrac{\partial u}{\partial x_i} \dfrac{\partial \Theta}{\partial x_i}$. It is easy to see that $F(x)$ has a compact support and $v \equiv u$ for $x_n > T + 1$.

First we construct a solution $w(x)$ of the equation

$$\Delta w = F(x)$$

in $S(-\infty, +\infty)$ such that $\frac{\partial u}{\partial \nu} = 0$ on $\sigma(-\infty, +\infty)$ and

$$\int\limits_{S(-\infty, +\infty)} (|w|^2 + |\nabla w|^2) e^{\alpha x_n} dx$$

$$\leqslant C \int\limits_{S(-\infty, +\infty)} |F|^2 e^{\alpha x_n} dx, \tag{52}$$

where $\alpha = const > 0$, C is a constant independent of α and w. For this purpose we write $F(x)$ in the form

$$F(x) = F_1(x) + F_2(x),$$

where

$$F_1(x) = F(x) - \frac{1}{\delta} \int\limits_{\omega(x_n)} F(x) dx', \quad F_2 = \frac{1}{\delta} \int\limits_{\omega(x_n)} F(x) dx'.$$

It is evident, that F_2 depends only on x_n, $\delta = \text{meas}_{n-1}\omega$, $\int\limits_{\omega(x_n)} F_1(x) dx' = 0$. Consider in $S(-\infty, +\infty)$ the equations

$$\Delta w_1 = F_1(x), \quad \Delta w_2 = F_2(x).$$

Let us construct w_1 in $S(-\infty, +\infty)$ satisfying the condition (52) and $\frac{\partial w_1}{\partial \nu} = 0$ on $\sigma(-\infty, +\infty)$. Consider w_1^N such that

$$\Delta w_1^N = F_1, \quad \text{in } S(-N, N), \tag{53}$$

$\frac{\partial w_1^N}{\partial \nu} = 0$ on $\sigma(-N, N)$, $w_1^N = 0$ at $x_n = N$ and $x_n = -N$, $N > 1$.

Multiplying equation (53) by $w_1^N e^{\alpha x_n}$, integrating it over $S(-N, N)$ and transforming the first term of the resulting equation by integration by parts, we get

$$-\int\limits_{S(-N,N)} |\nabla w_1^N|^2 e^{\alpha x_n} dx - \int\limits_{S(-N,N)} \frac{\partial w_1^N}{\partial x_n} w_1^N \alpha e^{\alpha x_n} dx$$

$$= \int\limits_{S(-N,N)} F_1 w_1^N e^{\alpha x_n} dx. \tag{54}$$

It follows from (54) that

$$\int\limits_{S(-N,N)} |\nabla w_1^N|^2 e^{\alpha x_n} dx$$

$$\leqslant \epsilon\alpha \int\limits_{S(-N,N)} |\nabla w_1^N|^2 e^{\alpha x_n} dx + \frac{1}{\epsilon}\alpha \int\limits_{S(-N,N)} |w_1^N|^2 e^{\alpha x_n} dx$$

$$+ \epsilon \int\limits_{S(-N,N)} |w_1^N|^2 e^{\alpha x_n} dx + \frac{1}{\epsilon} \int\limits_{S(-N,N)} |F_1|^2 e^{\alpha x_n} dx. \tag{55}$$

Let us prove that

$$\int\limits_{\omega(x_n)} w_1^N dx' = 0$$

for any $x_n > 0$. Integrating the equation $\Delta w_1^N = F_1$ over $S(N_1, N_2)$, we get

$$- \int\limits_{\omega(N_1)} \frac{\partial w_1^N}{\partial x_n} dx' + \int\limits_{\omega(N_2)} \frac{\partial w_1^N}{\partial x_n} dx'$$

$$= \int\limits_{S(N_1,N_2)} F_1 dx = 0, \quad (-N < N_1 < N_2 < N).$$

Therefore

$$\frac{\partial}{\partial x_n} \int\limits_{\omega(x_n)} w_1^N dx' = const, \quad \int\limits_{\omega(x_n)} w_1^N dx' = C_1 x_n + C_2, \ x_n > 0.$$

Since $w_1^N = 0$ for $x_n = -N$ and $x_n = N$, we get $C_1 = 0$, $C_2 = 0$. From the Poincaré inequality we get

$$\int\limits_{\omega(x_n)} |w_1^N|^2 dx' \leqslant C_3 \int\limits_{\omega(x_n)} |\nabla_{x'} w_1^N|^2 dx', \ x' = (x_1, \ldots, x_{n-1}). \tag{56}$$

From (55) and (56) it follows that if ϵ and α are sufficiently small,

$$\int\limits_{S(-N,N)} |w_1^N|^2 e^{\alpha x_n} dx + \int\limits_{S(-N,N)} |\nabla w_1^N|^2 e^{\alpha x_n} dx$$

$$\leqslant C_4 \int\limits_{S(-N,N)} |F_1|^2 e^{\alpha x_n} dx. \tag{57}$$

Therefore, there exists a sequence N such that $w_1^N \to w_1$ in $L^2(-a, a)$ for any $a > 0$ and w_1 is a solution of the equation (53) with the estimate

$$\int\limits_{S(-\infty,\infty)} |w_1|^2 e^{\alpha x_n} dx + \int\limits_{S(-\infty,\infty)} |\nabla w_1|^2 e^{\alpha x_n} dx$$

$$\leqslant C_4 \int\limits_{S(-\infty,\infty)} |F_1|^2 e^{\alpha x_n} dx \tag{58}$$

for sufficiently small $\alpha > 0$. Let us consider the equation

$$\Delta w_2 = F_2(x_n), \quad w_2'' = F_2(x_n), \tag{59}$$

in $S(-\infty, \infty)$. We apply to a solution of this equation the Hardy inequality
of the form

$$\int_{-\infty}^{+\infty} |f - C|^2 e^{\alpha x_n} \mathrm{d}x_n \leqslant \frac{4}{\alpha^2} \int_{-\infty}^{+\infty} |f'|^2 e^{\alpha x_n} \mathrm{d}x_n, \quad C, \alpha = const > 0.$$

This inequality follows from the classical Hardy inequality

$$\int_{0}^{\infty} t^{-2} |f|^2 \mathrm{d}t \leqslant 4 \int_{0}^{\infty} |f'|^2 \mathrm{d}t, \quad f(0) = 0,$$

upon changing the variable $t = e^{-\alpha x_n}$.

Then we have

$$\int_{-\infty}^{+\infty} e^{\alpha x_n} |w_2 - C_5 x_n - C_6|^2 \mathrm{d}x_n \leqslant \frac{4}{\alpha^2} \int_{-\infty}^{+\infty} |w_2' - C_5|^2 e^{\alpha x_n} \mathrm{d}x_n$$

$$\leqslant \frac{16}{\alpha^4} \int_{-\infty}^{+\infty} |w_2''|^2 e^{\alpha x_n} \mathrm{d}x_n. \qquad (60)$$

We shall consider a solution w_2 such that $C_5 = 0$, $C_6 = 0$. For $w = w_1 + w_2$ we have from (58), (60) that

$$\int_{S(-\infty,+\infty)} |\nabla w|^2 e^{\alpha x_n} + \int_{S(-\infty,+\infty)} |w|^2 e^{\alpha x_n} \mathrm{d}x$$

$$\leqslant C_7 \int_{S(-\infty,+\infty)} |F|^2 e^{\alpha x_n} \mathrm{d}x. \qquad (61)$$

From the theory of elliptic equations it is known that

$$\int_{S(k,k+1)} \sum_{|\beta| \leqslant 2} |D^\beta w|^2 \mathrm{d}x \leqslant C_8 \int_{S(k-1,k+2)} (|w|^2 + |\nabla w|^2 + |F|^2) \mathrm{d}x$$

for any $k > 0$. Multiplying this inequality by $e^{\alpha k}$ and summing over all
integers k, we obtain

$$\int_{S(-\infty,+\infty)} \sum_{|\beta| \leqslant 2} |D^\beta w|^2 e^{\alpha x_n} \mathrm{d}x$$

$$\leqslant C_9 \int\limits_{S(-\infty,+\infty)} (|w|^2 + |\nabla w|^2 + |F|^2)e^{\alpha x_n}\mathrm{d}x$$

$$\leqslant C_{10} \int\limits_{S(-\infty,+\infty)} |F|^2 e^{\alpha x_n}\mathrm{d}x. \tag{62}$$

Consider now the equation (51). By Theorem 7 we have $|g(x)| \leqslant C|x_n|^{-2-\epsilon}$ and we can suppose that $|g(x)| \leqslant \delta$ where δ is arbitrarily small since we can change a variable $\bar{x}_n = x_n - A$, where the constant A is sufficiently large.

We obtain a solution of equation (51) by successive approximations. For equations

$$\Delta v_N = g v_{N-1} + F(x), \quad N = 1, 2, \ldots, \tag{63}$$

we constructed a solution which satisfies the inequality (62). We take $v_0 \equiv 0$. The right side of (63) for any N has the finite integral

$$\int\limits_{S(-\infty,+\infty)} |g v_{N-1} + F(x)|^2 e^{\alpha x_n}\mathrm{d}x.$$

From (62) we have

$$\int\limits_{S(-\infty,+\infty)} \sum_{|\beta|\leqslant 2} |D^\beta v_N|^2 e^{\alpha x_n}\mathrm{d}x$$

$$\leqslant C_{11}\left[\int\limits_{S(-\infty,+\infty)} |F|^2 e^{\alpha x_n}\right.$$

$$\left. + \delta^2 \int\limits_{S(-\infty,+\infty)} \sum_{|\beta|\leqslant 2} |D^\beta v_{N-1}|^2 e^{\alpha x_n}\mathrm{d}x \right]. \tag{64}$$

From (63) we have

$$\Delta(v_N - v_{N-1}) = g(v_{N-1} - v_{N-2}) \text{ in } S(-\infty,+\infty)$$

and from (62) we get

$$\int\limits_{S(-\infty,+\infty)} \sum_{|\beta|\leqslant 2} |D^\beta(v_N - v_{N-1})|^2 e^{\alpha x_n}\mathrm{d}x$$

$$\leqslant \delta^2 C_{12} \int\limits_{S(-\infty,+\infty)} |v_{N-1} - v_{N-2}|^2 e^{\alpha x_n}\mathrm{d}x. \tag{65}$$

Since $C_7 \delta^2 < 1$, it follows from (65) that the sequence v_N converges in the norm $\|u\| = \left(\int\limits_{S(-\infty,+\infty)} |u|^2 e^{\alpha x_n} dx \right)^{\frac{1}{2}}$. From (62) we have

$$\int\limits_{S(-\infty,+\infty)} \sum_{|\beta| \leqslant 2} |D^\beta (v_N - v_{N+S})|^2 e^{\alpha x_n} dx$$

$$\leqslant \delta^2 C_{13} \int\limits_{S(-\infty,+\infty)} |v_{N-1} - v_{N+S-1}|^2 e^{\alpha x_n} dx \leqslant \epsilon,$$

where ϵ is arbitrarily small if N is sufficiently large. It means that $v_N \to \bar{v}$ as $N \to \infty$ and $\bar{v}$ satisfies equation (51).

From (64) we have

$$\int\limits_{S(-\infty,+\infty)} \sum_{|\beta| \leqslant 2} |D^\beta \bar{v}|^2 e^{\alpha x_n} dx \leqslant C_{14} \int\limits_{S(-\infty,+\infty)} |F|^2 e^{\alpha x_n} dx. \qquad (66)$$

Let us prove that $\bar{v} \to 0$ as $x_n \to \infty$. The function $\bar{v}$ satisfies equation (51) and therefore the inequality

$$\max_{S(\tau,\tau+1)} |\bar{v}| \leqslant C_{15} \int\limits_{S(\tau/2,\tau+2)} |\bar{v}|^2 dx \int\limits_{S(\tau/2,\tau+2)} |F|^2 dx$$

$$\leqslant C_{16} e^{-\alpha \tau/2} \left(\int\limits_{S(\tau/2,\tau+2)} |\bar{v}|^2 e^{\alpha x_n} dx + \int\limits_{S(\tau/2,\tau+2)} |F|^2 e^{\alpha x_n} dx \right)$$

$$\leqslant C_{17} e^{-\alpha \tau/2}$$

holds for $\tau > 0$ ([4],[5]).

We shall prove that $\bar{v} = v$. For this we need the following lemmas.

Lemma 5. *Let* $\Delta w = 0$ *in* $S(0,\infty)$, $\frac{\partial w}{\partial \nu} = 0$ *on* $\sigma(0,\infty)$, $w = 0$ *for* $x_n = 0$ *and for any* $x_n \geqslant 0$

$$\int\limits_{\omega(x_n)} w dx' = 0.$$

Then if $w \not\equiv 0$,

$$\int\limits_{S(0,t)} |\nabla w|^2 dx \geqslant C_1 e^{\gamma t}, \qquad (67)$$

$\gamma = const > 0$, $C_1 = const > 0$.

Proof. Integrating the equation $w\Delta w = 0$ over $S(0,t)$ and transforming the left hand side by integration by parts, we get

$$- \int\limits_{S(0,t)} |\nabla w|^2 dx + \int\limits_{\omega(t)} \frac{\partial w}{\partial x_n} w dx' = 0. \tag{68}$$

It follows from (68) and the Poincaré inequality that

$$\int\limits_{S(0,t)} |\nabla w|^2 dx \leqslant C_2 \int\limits_{\omega(t)} |\nabla w|^2 dx', \quad C_2 = const. \tag{69}$$

We set

$$J(t) = \int\limits_{S(0,t)} |\nabla w|^2 dx \ .$$

From (69) we have

$$J(t) \leqslant C_2 \frac{dJ}{dt}, \quad J(t) \geqslant J(t_1) e^{C_2(t-t_1)}, \quad t_1 = const > 0,$$

and therefore (67) is valid.

Lemma 6. Let $\Delta w = 0$ in $S(-\infty, 0)$, $\dfrac{\partial w}{\partial \nu} = 0$ on $\sigma(-\infty, 0)$ and

$$\int\limits_{S(-\infty,0)} |\nabla w|^2 e^{\alpha x_n} dx < \infty \tag{70}$$

for a sufficiently small positive constant α. Then

$$w(x', x_n) = ax_n + b + w_1(x),$$

where

$$\int\limits_{S(-\infty,0)} |w_1|^2 e^{-\alpha x_n} dx < \infty, \quad a, b = const. \tag{71}$$

Proof. Let $w(x) = w_1(x) + w_2(x_n)$, where

$$\int\limits_{\omega(x_n)} w_1(x) dx' = 0.$$

We have

$$
\begin{aligned}
w_2''(x_n) &= \frac{1}{\text{meas } \omega} \int\limits_{\omega(x_n)} \frac{d^2 w}{dx_n^2} dx' = -\frac{1}{\text{meas } \omega} \int\limits_{\omega(x_n)} \Delta_{x'} w_2 dx' \\
&= -\frac{1}{\text{meas } \omega} \int\limits_{\partial \omega} \frac{\partial w_2}{\partial \nu} d_s = 0.
\end{aligned}
$$

Therefore

$$
w_2(x_n) = a x_n + b, \quad a, b = const, \quad \Delta w_1 = 0 \text{ in } S(-\infty, 0).
$$

Let us prove (71). Consider a sequence of functions $V_N(x)$ such that

$$
V_N(x) = w_1 \text{ for } x_n = 0, \quad \frac{\partial V_N}{\partial \nu} = 0 \text{ on } \sigma(-N, 0),
$$
$$
\Delta V_N(x) = 0 \text{ in } S(-N, 0), \quad V_N = 0 \text{ for } x_n = -N.
$$

It is easy to see that

$$
\frac{\partial}{\partial x_n} \int\limits_{\omega(x_n)} V_N dx' = const. \quad \int\limits_{\omega(x_n)} V_N dx' = a_1(x_n + N) + b_1.
$$

Since $V_N = 0$ for $x_n = -N$ and according to the maximum principle V_N is bounded by a constant which does not depend on N, we have $b_1 = 0$ and $a_1 = 0$. Considering the function $W_N = V_N - w_1 \chi(x_n)$, where $\chi(x_n) = 1$ for $x_n > -1$ and $\chi(x_n) = 0$ for $x_n < -2$, $\chi \in C^\infty(\mathbf{R}^1)$, we get $\Delta W_N = -\Delta(w_1 \chi(x_n))$, $W_N = 0$ for $x_n = 0$ and for $x_n = -N$, and we have

$$
\int\limits_{S(-N,0)} |\nabla W_N|^2 e^{-\alpha x_n} dx + \int\limits_{S(-N,0)} \frac{\partial W_N}{\partial x_n} W_N \alpha e^{-\alpha x_n} dx
$$
$$
= -\int\limits_{S(-N,0)} \Delta(w_1 \chi(x_n)) W_N e^{-\alpha x_n} dx,
$$
$$
\int\limits_{S(-N,0)} |\nabla W_N|^2 e^{-\alpha x_n} dx
$$
$$
\leqslant \alpha\epsilon \int\limits_{S(-N,0)} |\nabla W_N|^2 e^{-\alpha x_n} dx + \alpha \frac{1}{\epsilon} \int\limits_{S(-N,0)} |W_N|^2 e^{-\alpha x_n} dx + M_1,
$$

where M_1 is a constant which does not depend on N. Using the Poincaré inequality to estimate

$$\int\limits_{S(-N,-2)} |W_N|^2 e^{-\alpha x_n} dx,$$

since $W_N = V_N$ for $x_n < -2$ and taking α sufficiently small, we get

$$\int\limits_{S(-N,0)} |\nabla W_N|^2 e^{-\alpha x_n} dx < \infty. \qquad (72)$$

From (72) and the boundedness of the functions V_N it follows that a sub-sequence of V_N converges to V, as $N \to -\infty$, uniformly on any compact subset of $S(-\infty, 0)$, $\Delta V = 0$ in $S(-\infty, 0)$, $V = w_1$ for $x_n = 0$ and

$$\int\limits_{S(-\infty,0)} |\nabla V|^2 e^{-\alpha x_n} dx < \infty \qquad (73)$$

for all sufficiently small $\alpha > 0$.

Consider now $W = w_1 - V$. We have $\Delta W = 0$ in $S(-\infty, 0)$, $W|_{x_n=0} = 0$ and

$$\int\limits_{\omega(x_n)} W dx' = 0.$$

According to Lemma 5, if $W \not\equiv 0$, then

$$\int\limits_{S(-t,0)} |\nabla W|^2 dx \geqslant C_1 e^{\gamma t}, \quad C_1, \gamma = const > 0. \qquad (74)$$

From (70) and (73) it follows that

$$\int\limits_{S(-\infty,0)} |\nabla W| e^{-\alpha x_n} dx < \infty,$$

$$\int\limits_{S(-N,0)} |\nabla W|^2 dx \leqslant \sum_{i=0}^{N} \int\limits_{S(-i,-i+1)} |\nabla W|^2 e^{\alpha x_n} e^{-\alpha x_n} dx \leqslant C_2 e^{\alpha N}$$

for all sufficiently small α. This contradicts (74). Therefore, $W \equiv 0$ and $w_1 = V$. Since $\int\limits_{\omega(x_n)} w_1 dx' = 0$, the inequality (71) follows from (73) and the Poincaré inequality. The lemma is proved.

Now we shall prove that $\bar{v} = v$ in $S(-\infty, +\infty)$. We have for $W = \bar{v} - v$ the equation

$$\Delta W - gW = 0 \text{ in } S(-\infty, +\infty), \quad \frac{\partial W}{\partial \nu} = 0 \text{ on } \sigma(-\infty, +\infty)$$

and $W \to 0$ as $x_n \to \infty$, since $\bar{v} \to 0$ and $v \to 0$ as $x_n \to \infty$. For $x_n < 0$ we have $\Delta W = 0$ and $v \equiv 0$. For $\bar{v}$ the inequality (66) is satisfied. According to Lemma 6

$$W(x) = ax_n + b + w_1(x), \quad a, b = const, \qquad (75)$$

where for the function $w_1(x)$ the inequality (71) holds. Let us prove that $a = 0$, $b = 0$. Then we have that $W = w_1$ for $x_n < 0$. Because of (71) $w_1(x) \to 0$ as $x_n \to -\infty$ and according to the maximum principle $W \equiv 0$ in $S(-\infty, \infty)$, $\bar{v} = v$ and for $\bar{v}$ the estimate (66) is valid.

If $a = 0$ in (75), then $W = b + w_1(x)$. We have

$$-\int\limits_{S(-N,N)} |\nabla W|^2 dx - \int\limits_{S(-N,N)} g|W|^2 dx + \int\limits_{\omega(N)} W\frac{\partial W}{\partial \nu} dx'$$

$$-\int\limits_{\omega(-N)} W\frac{\partial W}{\partial \nu} dx' = 0. \qquad (76)$$

When $N \to \infty$, the third integral in (76) tends to zero, since $W \to 0$ and $\frac{\partial W}{\partial x_i} \to 0$, $i = 1, \dots, n$, as $x_n \to \infty$. The last integral in (76) also tends to zero as $N \to \infty$. This integral can be written in the form

$$\int\limits_{\omega(-N)} (b + w_1)\frac{\partial w_1}{\partial \nu} dx'.$$

From (71) it follows that w_1 is bounded in $S(-\infty, 0)$, $\frac{\partial w_1}{\partial x_i} \to 0$, $i = 1, \dots, n$, as $x_n \to -\infty$. Since $g \geqslant 0$, from (76) we get that $\text{grad}W = 0$ in $S(-\infty, +\infty)$. Taking into account that $W \to 0$ as $x_n \to \infty$, we conclude that $W \equiv 0$, $v = \bar{v}$ in $S(-\infty, +\infty)$.

If we suppose that $a \neq 0$, we get a contradiction. Indeed, suppose that $a < 0$. This means that W preserves the sign. We prove that if W preserves the sign, then $W \geqslant C_0 = const > 0$ in $S(-\infty, +\infty)$. This is impossible, since $W \to 0$ as $x_n \to \infty$. In order to prove that $W \geqslant C_0 > 0$ we construct the sequence of functions $y_N(x_n)$ such that

$$y_N'' - \frac{C}{1 + |x_n|^{2+\epsilon}} y_N = 0 \text{ in } S(l, N),$$

where $C = const > 0$, the constant l chosen in such a way that $W > 1$ for $x_n = l$ and

$$y_N(l) = 1, \quad y_N(N) = 0.$$

According to the maximum principle, $0 \leqslant y_N \leqslant 1$, we have $y_N'' \geqslant 0$,

$$y_N'(x_n) = y_N'(N) - \int_{x_n}^{N} y_N'' dx_n < 0, \text{ since } y'(N) < 0, \ y_N'(x_n) = y_N'(\xi) +$$

$\int_{\xi}^{x_n} y_N'' dx_n$. The last equality means that the $|y_N'(x_n)|$ are bounded with respect to N on any finite interval (l, ∞) of x_n. Therefore, y_N converges to y uniformly on any finite interval of x_n and for y we have

$$y''(x_n) - \frac{C}{1 + |x_n|^{2+\epsilon}} y(x_n) = 0, \tag{77}$$

$y' < 0$ for $l < x_n < +\infty$ and $y \to C_0$ as $x_n \to \infty$. It is easy to see that $W > y$. For $W - y_N = z_N$ we have the equation

$$\Delta z_N - g(x)z_N = gy_N - \frac{C}{1 + |x_n|^{2+\epsilon}} y_N \leqslant 0, \tag{78}$$

since the constant C is chosen in such a way that $gy_N - \frac{C}{1 + |x_n|^{2+\epsilon}} y_N < 0$, and $z_N(l) \geqslant 0$, $z_N(N) > 0$. From (78) it follows that z cannot attain a negative value in the interval (l, N). Therefore, $z_N \geqslant 0$, $W \geqslant y_N$ and $W \geqslant y$ for $x_n > l$. According to the Bellman theorem [6] any solution of the equation

$$y'' + Q(t)y = 0, \quad \text{on } [0, \infty], \quad y \not\equiv 0,$$

where

$$\int_0^\infty |t| |Q(t)| dt < \infty,$$

has the form

$$y(t) = a_1 t + a_2 + b(t)$$

and $b(t) \to 0$ as $t \to \infty$, $a_1^2 + a_2^2 \neq 0$.

It follows from this theorem that $C_0 \neq 0$. But $W > y$ and $W \to 0$ as $x_n \to \infty$. This contradiction proves that $a = 0$ and $W = 0$, $v = \bar{v}$ in $S(-\infty, +\infty)$. The theorem is proved.

The case when a solution of the equation (23) in $S(0, \infty)$ with the boundary condition $\dfrac{\partial u}{\partial \nu} = 0$ on $\sigma(0, \infty)$ preserves sign will be considered in more detail in these lectures in section 3.

Here we prove a particular result.

Theorem 9. *Assume that $u(x) \geqslant 0$, $u \not\equiv 0$ in $S(0, \infty)$, $\Delta u - |u|^{p-1} u = 0$ and $\dfrac{\partial u}{\partial \nu} = 0$ on $\sigma(0, \infty)$, $p > 1$. Then there exists a constant $C_1 > 0$ such that*

$$u(x', x_n) \geqslant C_p (x_n + C_1)^{\frac{2}{1-p}}, \quad C_p = \left(\frac{2(1+p)}{(p-1)^2} \right)^{\frac{1}{p-1}}.$$

Proof. We choose the constant C_1 in such a way that

$$C_p (A + C_1)^{\frac{2}{1-p}} < \min_{x' \in \omega} u(x', A), \quad A = const > 0.$$

Then we consider the function $w = C_p (x_n + C_1)^{\frac{2}{1-p}} - u$ and, using the maximum principle and the Lemma on the normal derivative ([2],[3]) we prove that $w \leqslant 0$ in $S(A, \infty)$.

Remark 1. From Theorems 3 and 9 for positive solutions $u(x)$ of the equation (23) with $k = 1$ and $\dfrac{\partial u}{\partial \nu} = 0$ on $\sigma(0, \infty)$ we have

$$u(x', x_n) = C_p x^{\frac{2}{1-p}} + \mathcal{O} \left(x_n^{\frac{2}{1-p} - 1} \right) \quad \text{as } x_n \to \infty.$$

Some additional references on the subject of this lecture one can find at the end of Chapter 1.

References

[1] H. Berestycki and L. Nirenberg, Some qualitative properties of solutions of semi-linear elliptic equations in cylindrical domains, *Analysis*, ed. by P. Rabinowitz, Academic Press, 1990, 115–164.

[2] C. Miranda, Equazioni alle derivate parziali di tipo ellittico, *Ergebn. Math.*, Berlin, 1955.

[3] O.A. Oleinik, On properties of solutions of some boundary value problems for equations of elliptic type,*Math. Sbornik*, **30**,1952, 695–702.

[4] E. De Giorgi, Sulla differenziabilità e l'analiticità delle estremali degli integrali, *Mem. Acc. Sci. Torino*, 1957, 1–19.

[5] D. Gilbarg, N.S. Trudinger, Elliptic partial differential equations of second order, Springer Verlag, 1983.

[6] R. Bellman, Stability theory of differential equations, McGraw-Hill, New York, 1953.

2.2 On the asymptotics at infinity of solutions in cylindrical domains for a class of elliptic second order equations containing first derivatives.

In this lecture we consider nonlinear elliptic equations of the form

$$\sum_{i,j=1}^{n} \frac{\partial}{\partial x_i}\left(a_{ij}(x')\frac{\partial u}{\partial x_j}\right) + \sum_{j=1}^{n-1} a_j(x')\frac{\partial u}{\partial x_j} + K\frac{\partial u}{\partial x_n} - a_0|u|^{p-1}u = 0, \quad (1)$$

where $x' = (x_1, \ldots, x_{n-1})$, $x = (x', x_n)$, $x' \in \omega$, ω is a bounded smooth domain in $\mathbf{R}^{n-1}$, $K = const$, $a_0 = const > 0$ and $p > 1$. We will use here the same notation as in section 1. We set

$$S(a,b) = \{x : x' \in \omega, a < x_n < b\}, \qquad \sigma(a,b) = \{x : x' \in \partial\omega, a < x_n < b\}.$$

We assume that

$$\lambda_1|\xi|^2 \leqslant \sum_{i,j=1}^{n} a_{ij}(x')\xi_i\xi_j \leqslant \lambda_2|\xi|^2, \ \forall \xi \in \mathbf{R}^n, \ \lambda_1, \lambda_2 = const > 0,$$

where the coefficients $a_{ij}(x')$, $a_i(x')$ are bounded measurable functions and $a_{ij} = a_{ji}$.

We study the boundary value problem for the equation (1) in $S(0,\infty)$ with the boundary condition

$$\frac{\partial u}{\partial \mu} = 0 \text{ on } \sigma(0,\infty), \quad \frac{\partial u}{\partial \mu} \equiv \sum_{i,j=1}^{n} a_{ij}\frac{\partial u}{\partial x_j}\nu_i, \quad (2)$$

where $\nu = (\nu_1, \ldots, \nu_n)$ is a normal direction to $\sigma(0,\infty)$.

The equations of this form are important for the theory of travelling waves, for boundary layer theory and other problems of mathematical physics. They are the subject of a vast literature (see, for example, [1]–[2] and references there). Our aim is to study the behavior of the solutions of problem (1),(2) when $x_n \to \infty$.

We assume that

$$a_{nn} = 1, \quad a_{nj} = 0, \text{ for } j < n.$$

For the case when the $\dfrac{\partial a_{ij}}{\partial x_i}$ are bounded, it was proved in section 1 that $u(x', x_n) \to 0$ as $x_n \to \infty$.

We define a weak solution of (1),(2) as a function $u(x)$ which belongs to $H^1(S(0, M))$ for any $M < \infty$, is bounded in $S(0, M)$ and satisfies the integral identity

$$
-\int\limits_{S(0,\infty)} \sum_{i,j=1}^{n} a_{ij}(x') \frac{\partial u}{\partial x_j} \frac{\partial \varphi}{\partial x_i} dx + \sum_{i=1}^{n-1} \int\limits_{S(0,\infty)} a_i(x') \frac{\partial u}{\partial x_i} \varphi dx
$$

$$
+ K \int\limits_{S(0,\infty)} \frac{\partial u}{\partial x_n} \varphi(x) dx - a_0 \int\limits_{S(0,\infty)} |u|^{p-1} u \varphi dx = 0 \tag{3}
$$

for any function $\varphi(x)$ which belongs to $H^1(S(0,\infty))$ and is equal to zero for $x_n \geqslant X_n$, where X_n depends on φ.

In what follows we will use the following known results, which one can get easily from [3]–[6].

Proposition 1. *Consider the equation*

$$
\sum_{i,j=1}^{n} \frac{\partial}{\partial x_i} \left(a_{ij}(x') \frac{\partial u}{\partial x_j} \right) + \sum_{i=1}^{n-1} a_i(x') \frac{\partial u}{\partial x_i} + a_n(x') \frac{\partial u}{\partial x_n} - a_0(x) u
$$

$$
= F_0 + \sum_{i=1}^{n} \frac{\partial F_i}{\partial x_i}. \tag{4}
$$

Assume that

(a) $J_h(F) \equiv \int\limits_{S(-\infty,+\infty)} \left(|F_0|^2 e^{2hx_n} + \sum_{i=1}^{n} |F_i|^2 e^{2hx_n} \right) dx < \infty,$
    $h = const.$

(b) *The eigenvalue problem*

$$
-\lambda^2 u + \sum_{i,j=1}^{n-1} \frac{\partial}{\partial x_i} \left(a_{ij}(x') \frac{\partial u}{\partial x_j} \right) + \sum_{i=1}^{n-1} a_i(x') \frac{\partial u}{\partial x_i} + i\lambda a_n(x') u = 0 \quad in \ \omega,
$$

$$
\tag{5}
$$

$$
\frac{\partial u}{\partial \mu} = 0 \quad on \ \partial \omega, \tag{6}
$$

does not have spectral points λ with $\operatorname{Im} \lambda = h$.

Then there exists $\epsilon > 0$ such that if $|a_0(x)| \leqslant \epsilon$, $|a_n(x)| < \epsilon$, the unique solution of equation (4) in $S(-\infty,+\infty)$ with the boundary condition

$$
\frac{\partial u}{\partial \mu} = 0 \quad on \ \sigma(-\infty,+\infty) \tag{7}
$$

exists and

$$I_h(u) \equiv \int\limits_{S(-\infty,+\infty)} \left[\sum_{i=1}^n \left(\frac{\partial u}{\partial x_i} \right)^2 e^{2hx_n} + |u|^2 e^{2hx_n} \right] dx \leqslant C J_h(F). \quad (8)$$

Proposition 2. *Consider the equation*

$$\sum_{i,j=1}^n \frac{\partial}{\partial x_i} \left(a_{ij}(x') \frac{\partial u}{\partial x_j} \right) + \sum_{i=1}^{n-1} a_i(x') \frac{\partial u}{\partial x_i} + a_n(x') \frac{\partial u}{\partial x_n} = F + \sum_{i=1}^n \frac{\partial F_i}{\partial x_i} \quad (9)$$

and make the following assumptions:

(a) $J_{h_1}(F) + J_{h_2}(F) < \infty$.

(b) On the straight line $\operatorname{Im} \lambda = h_2$ there is no spectrum of the eigenvalue problem (5),(6).

(c) $I_{h_1}(u) < \infty$.

Then for a solution $u(x)$ of the equation (9) in $S(-\infty,+\infty)$ with the boundary condition (7) on $\sigma(-\infty,+\infty)$ the representation

$$u(x) = \sum_{\substack{h_2 < \operatorname{Im} \lambda_j < h_1 \\ 0 \leqslant k \leqslant k(j)}} C_{jk} x_n^{k(j)-k} \phi_{jk}(x') e^{i\lambda_j x_n} + u_1(x), \quad (10)$$

holds where C_{jk} are constants, λ_j belongs to the spectrum of the problem (5),(6), $\phi_{j0}, \ldots, \phi_{jk(j)}$ is a chain of eigenfunctions and adjoint functions corresponding to the eigenvalue λ_j. Furthermore

$$I_{h_2}(u_1) \leqslant C_1 J_{h_1}(F) + C_2 J_{h_2}(F), \quad C_1, C_2 = const. \quad (11)$$

Theorem 1. *Let $u(x)$ be a solution of the problem (1),(2) such that $u(x)$ has positive and negative values in $S(m,\infty)$ for any $m > 0$ (oscillating). Then*

$$|u(x)| \leqslant C e^{-\alpha x_n}, \quad in \ S(0,+\infty), \quad (12)$$

where C, $\alpha = const > 0$ and α does not depend on u.

Proof. As proved in section 1,

$$u(x) \to 0, \quad as \ x_n \to \infty.$$

If X is sufficiently large, then $|a_0(x)| = a_0|u|^{p-1} < \epsilon$ for $x_n > X$. Consider the function $v(x) = \theta(x_n)u$, where $\theta(x_n) = 0$ for $x_n < X$, $\theta(x) = 1$ for $x_n > X + 1$, $\theta \in C^\infty(\mathbf{R}^1)$. The function $v(x)$ satisfies in $S(-\infty, +\infty)$ the equation

$$\sum_{i,j=1}^n \frac{\partial}{\partial x_i}\left(a_{ij}(x')\frac{\partial v}{\partial x_j}\right) + \sum_{i=1}^{n-1} a_i(x')\frac{\partial v}{\partial x_i} + K\frac{\partial v}{\partial x_n} - a_0(x)v$$

$$= F_0 + \sum_{i=1}^n \frac{\partial F_i}{\partial x_i} \tag{13}$$

with the boundary condition $\dfrac{\partial v}{\partial \mu} = 0$ on $\sigma(-\infty, \infty)$, where F_0, F_i belong to $L_2(S(-\infty, +\infty))$ and have compact support with respect to x_n.

We can assume that $a_0(x) = 0$ for $x_n < X$. It is evident that

$$\frac{\partial v}{\partial \mu} = 0 \ \text{ on } \sigma(-\infty, \infty). \tag{14}$$

Consider the eigenvalue problem

$$\sum_{i,j=1}^{n-1} \frac{\partial}{\partial x_i}\left(a_{ij}(x')\frac{\partial u}{\partial x_j}\right) + \sum_{i=1}^{n-1} a_i(x')\frac{\partial u}{\partial x_i} + iK\lambda u - \lambda^2 u = 0, \quad x \in \omega, \tag{15}$$

$$\frac{\partial u}{\partial \mu} = 0 \ \text{ on } \partial\omega. \tag{16}$$

It is clear that $\lambda = 0$ is an eigenvalue of the eigenvalue problem (15),(16) and $u = const$ is an eigenfunction, corresponding to $\lambda = 0$.

Let us prove that on the straight line $\operatorname{Im}\lambda = 0$ in the complex plane λ there is only one spectral point $\lambda = 0$ of the eigenvalue problem (15), (16). Indeed, let $\lambda = s \in \mathbf{R}^1$ be a spectral point of the problem (15),(16) and $s \neq 0$.

The function $\phi_1(x) = e^{isx_n}\phi(x')$, where $\phi(x')$ is an eigenfunction of the problem (15),(16) corresponding to the eigenvalue $\lambda = s$, satisfies the equation

$$\sum_{i,j=1}^n \frac{\partial}{\partial x_i}\left(a_{ij}(x')\frac{\partial \phi_1}{\partial x_j}\right) + \sum_{i=1}^n a_i(x')\frac{\partial \phi_1}{\partial x_i} = 0, \quad a_n = K, \tag{17}$$

in $S(-\infty, \infty)$ and the boundary condition

$$\frac{\partial \phi_1}{\partial \mu} = 0 \ \text{ on } \sigma(-\infty, \infty). \tag{18}$$

Let

$$\phi(x') = \psi_1(x') + i\psi_2(x'),$$

where $\psi_1(x')$, $\psi_2(x')$ are the real and imaginary parts of $\phi(x')$ respectively. In this case, the function

$$\phi_2(x) = \cos(sx_n)\psi_1(x') - \sin(sx_n)\psi_2(x')$$

is a solution of the equation (17) in $S(-\infty,\infty)$ with the boundary condition (18). The function $\phi_2(x)$ is a periodic function with respect to x_n and according to the maximum principle (see [7]) $\phi_2(x) = const.$ It is easy to see that this is possible, if and only if $s = 0$.

Thus, the eigenvalue problem (15),(16) has a unique spectral point $\lambda = 0$ for which $\operatorname{Im}\lambda = 0$.

If $K = 0$, then for $\lambda = 0$ we have the eigenfunction $u_1 \equiv 1$ and the adjoint function $u_2 \equiv 1$ for the eigenvalue problem (15),(16). Indeed, suppose that we have the second adjoint function $u_3(x)$. Then we have equations

$$\sum_{i,j=1}^{n-1} \frac{\partial}{\partial x_i}\left(a_{ij}(x')\frac{\partial u_1}{\partial x_j}\right) + \sum_{i=1}^{n-1} a_i(x')\frac{\partial u_1}{\partial x_i} = 0 \quad \text{in } \omega,$$

$$\frac{\partial u_1}{\partial \mu} = 0 \quad \text{on } \partial\omega,$$

$$\sum_{i,j=1}^{n-1} \frac{\partial}{\partial x_i}\left(a_{ij}(x')\frac{\partial u_2}{\partial x_j}\right) + \sum_{i=1}^{n-1} a_i(x')\frac{\partial u_2}{\partial x_i} = 0 \quad \text{in } \omega,$$

$$\frac{\partial u_2}{\partial \mu} = 0 \quad \text{on } \partial\omega,$$

and

$$\sum_{i,j=1}^{n-1} \frac{\partial}{\partial x_i}\left(a_{ij}(x')\frac{\partial u_3}{\partial x_j}\right) + \sum_{i=1}^{n-1} a_i(x')\frac{\partial u_3}{\partial x_i} - 1 = 0 \quad \text{in } \omega, \qquad (19)$$

$$\frac{\partial u_3}{\partial \mu} = 0 \quad \text{on } \partial\omega. \qquad (20)$$

If the solution $u_3(x)$ of the problem (19),(20) exists, then the function

$$v = \frac{x_n^2}{2} - u_3(x')$$

is a solution of the equation

$$\sum_{i,j=1}^{n} \frac{\partial}{\partial x_i}\left(a_{ij}(x')\frac{\partial v}{\partial x_j}\right) + \sum_{i=1}^{n-1} a_i(x')\frac{\partial v}{\partial x_i} = 0 \ \text{in} \ S(-\infty,\infty), \qquad (21)$$

with the boundary condition

$$\frac{\partial v}{\partial \mu} = 0 \ \text{on} \ \sigma(-\infty,\infty). \qquad (22)$$

The existence of such a solution v contradicts the maximum principle, since $v(x)$ takes a minimum value inside $S(-\infty,\infty)$.

In the case $K \neq 0$ in the equation (1), $\lambda = 0$ is also the only spectral point with $\text{Im}\,\lambda = 0$ for the problem (15),(16) with the eigenfunction $u_1 \equiv 1$ and there is no adjoint function. Indeed, if $u_1 \equiv 1$ is the eigenfunction and $u_2(x')$ is an adjoint function, then by the definition of an adjoint function we have

$$\sum_{i,j=1}^{n-1} \frac{\partial}{\partial x_i}\left(a_{ij}(x')\frac{\partial u_2}{\partial x_j}\right) + \sum_{i=1}^{n-1} a_i(x')\frac{\partial u_2}{\partial x_i} + \mathrm{i}K = 0 \ \text{in} \ \omega,$$

$$\frac{\partial u_2}{\partial \mu} = 0 \ \text{on} \ \partial\omega.$$

In this case

$$v(x) = \frac{x_n^2}{2} + \frac{\text{Im}\,u_2(x')}{K}$$

is the solution of equation (21) with the boundary condition (22). This is impossible due to the maximum principle, since $v(x)$ attains its minimum value inside $S(-\infty,\infty)$.

It is well-known that the spectrum of the eigenvalue problem (15),(16) is discrete [8]. It is easy to prove that in any strip $|\text{Im}\,\lambda| < H$ of the complex plane λ there are no spectral points, if $|\text{Re}\,\lambda| > A$ and A is sufficiently large. Indeed, multiplying the equation (15) by $\bar{u}$, transforming the first integral by integration by parts and taking into account the boundary condition (16), we obtain the inequality

$$\int_\omega \sum_{i=1}^{n-1} |u_{x_i}|^2 \mathrm{d}x'$$

$$\leqslant \epsilon \sum_{i=1}^{n-1} \int_\omega |u_{x_i}|^2 \mathrm{d}x' + C_1 \int_\omega |u|^2 \mathrm{d}x' + H^2 \int_\omega |u|^2 \mathrm{d}x'$$

$$-\frac{A^2}{2}\int_\omega |u|^2 \mathrm{d}x', \quad \epsilon = const, \tag{23}$$

if A is sufficiently large. But since ϵ is an arbitrarily small number, the inequality (23) for sufficiently large values of A is possible only in the case $u \equiv 0$.

Because of the above property of the spectrum of the problem (15),(16) and the discreteness of this spectrum we can conclude that there exists $\epsilon_1 = const > 0$ such that in the strip $|\operatorname{Im}\lambda| < \epsilon_1$ there is only one spectral point $\lambda = 0$ of the eigenvalue problem (15), (16).

From Proposition 1, formulated above, it follows that there exists a solution w of the problem (13) with the boundary condition $\dfrac{\partial w}{\partial \mu} = 0$ on $\sigma(-\infty, \infty)$ such that

$$I_{\epsilon_1}(w) < \infty. \tag{24}$$

Consider the function

$$Z(x) = w(x) - v(x). \tag{25}$$

From (24) and the estimate of the De Giorgi type it follows that $|w(x)| \leqslant C_1 e^{-\epsilon_1 x_n}$ for $x_n > 0$ and therefore $Z(x) \to 0$ as $x_n \to \infty$. Indeed, from (24) we have

$$\int_{S(T-1,T+2)} |w|^2 e^{2\epsilon_1 x_n} \mathrm{d}x < \infty.$$

According to the De Giorgi theorem ([9], [7]) for $T > X + 2$

$$\max_{S(T,T+1)} |w|^2$$
$$\leqslant C_2 \int_{S(T-1,T+2)} |w|^2 \mathrm{d}x$$
$$\leqslant C_2 e^{-2\epsilon_1(T-1)} \int_{S(T-1,T+2)} e^{2\epsilon_1 x_n} |w|^2 \mathrm{d}x$$
$$\leqslant C_3 e^{-2\epsilon_1 T}. \tag{26}$$

In order to use Proposition 2 let us consider the function

$$Z_1(x) = Z(x)\theta_1(x_n),$$

where $\theta_1(x_n) = 1$ for $x_n < X - 1$ and $\theta_1(x_n) = 0$ for $x_n > X$, $\theta_1(x_n) \in C^\infty(\mathbf{R}^1)$. The function $Z_1(x)$ satisfies the equation of the form (9) and one can apply Proposition 2 to $Z_1(x)$, since the coefficient of Z_1 in this equation is equal to zero.

From Proposition 2 it follows that for $x_n < X - 1$

$$Z_1(x) = Z(x) = w(x) = C_4 + C_5 x_n + \mathcal{O}\left(e^{\epsilon_1 x_n}\right) \quad \text{as } x_n \to -\infty. \quad (27)$$

If $C_4 = C_5 = 0$, then $Z(x) \equiv 0$ by the maximum principle and because of $Z(x) \to 0$ as $x_n \to +\infty$ and $x_n \to -\infty$. In this case $v(x) = w(x)$ and the assertion of Theorem 1 follows from (26).

If one of the constants C_4 and C_5 is not equal to zero, the maximum principle implies that $Z(x)$ preserves sign in $S(-\infty, \infty)$. Suppose that $Z(x) > 0$ in $S(-\infty, \infty)$. If $Z(x)$ takes negative values, we consider $-Z(x)$. So we have

$$u(x) \leqslant w(x) \text{ in } S(X + 1, \infty), \text{ and } u(x) \leqslant C_6 e^{-\epsilon_1 x_n} \text{ in } S(X + 1, +\infty). \quad (28)$$

Let us prove that

$$u(x) \geqslant -C_7 e^{-\epsilon_1 x_n}.$$

For this, consider the function

$$V(x) = -u + 2C_6 e^{-\epsilon_1 T}. \quad (29)$$

It is evident that $V > 0$ in $S(T, T + 1)$ for $T > X + 1$ and V satisfies the equation

$$\sum_{i,j=1}^{n} \frac{\partial}{\partial x_i}\left(a_{ij}(x')\frac{\partial V}{\partial x_j}\right) + \sum_{i=1}^{n-1} a_i(x')\frac{\partial V}{\partial x_i} + K\frac{\partial V}{\partial x_n} - a_0(x)V$$
$$= -2a_0(x)C_6 e^{-\epsilon_1 T} \quad (30)$$

in $S(T, T + 1)$ and the boundary condition

$$\frac{\partial V}{\partial \mu} = 0 \quad \text{on } \sigma(T, T + 1). \quad (31)$$

Consider the solution $V_1(x)$ of the equation (30) in $S(T, T + 1)$ such that

$$V_1 = 0 \text{ for } x_n = T, \quad V_1 = 0 \text{ for } X_n = T + 1, \quad \frac{\partial V_1}{\partial \mu} = 0 \text{ on } S(T, T + 1).$$

From the estimates for linear equations in the $L_2(T, T + 1)$ norm and the uniqueness of V_1 we have [7]

$$|V_1(x)| \leqslant C_8 \max |a_0(x)| e^{-\epsilon_1 T} \leqslant \frac{C_6}{2} e^{-\epsilon_1 T}, \quad (32)$$

if T is sufficiently large, since $a_0(x) \to 0$ as $x_n \to \infty$. Let $W = V - V_1$. From (28), (29), (32) it follows that $W > 0$ in $S(T, T+1)$. The function $u(x)$ changes the sign in $S(T, T+1)$. Therefore, there is a point $x_T \in S(T+1/4, T+1/2)$ such that $u(x_T) = 0$. From (28), (29), (32) it follows that

$$W(x_T) < C_6 e^{-\epsilon_1 T}.$$

From the Harnack inequality we have

$$W(x) \leqslant C_9 e^{-\epsilon_1 T} \text{ in } S(T+1/4, T+1/2).$$

Therefore, $-u(x) + 2C_6 e^{-\epsilon_1 T} - \frac{C_4}{2} e^{-\epsilon_1 T} \leqslant C_9 e^{-\epsilon_1 T}$ and

$$u(x) \geqslant -C_{10} e^{-\epsilon_1 T} \text{ in } S(T+1/4, T+1/2).$$

We have proved that

$$|u(x)| \leqslant C e^{-\epsilon_1 x_n}, \quad C = const,$$

and this completes the proof of the theorem.

Theorem 2. *If $K > 0$, then any solution of the problem (1),(2) satisfies the inequality*

$$|u(x)| \leqslant C_1 e^{-\delta x_n}, \quad C_1, \delta = const > 0. \tag{33}$$

Proof. Calculations show that for $u_1(x_n) = C_1 e^{-\delta x_n}$

$$\sum_{i,j=1}^{n} \frac{\partial}{\partial x_i} \left(a_{ij}(x') \frac{\partial u_1}{\partial x_j} \right) + \sum_{i=1}^{n-1} a_i(x') \frac{\partial u_1}{\partial x_i} + K \frac{\partial u_1}{\partial x_n} - a_0 |u|^{p-1} u_1$$
$$= (\delta^2 - K\delta) C_1 e^{-\delta x_n} - a_0 |u|^{p-1} C_1 e^{-\delta x_n} < 0 \text{ in } S(0, \infty),$$

when $K > \delta$ and $C_1 > 0$ is any constant. Therefore for $V = C_1 e^{-\delta x_n} - u(x)$ we have

$$\sum_{i,j=1}^{n} \frac{\partial}{\partial x_i} \left(a_{ij}(x') \frac{\partial V}{\partial x_j} \right) + \sum_{i=1}^{n-1} a_i(x') \frac{\partial V}{\partial x_i} + K \frac{\partial V}{\partial x_n} - Q(x)V \leqslant 0,$$

where $Q(x) = a_0 |u|^{p-1}$. If $C_1 \geqslant \max_{x_n=0} |u|$, then from the maximum principle it follows that $V \geqslant 0$ in $S(0, \infty)$ and therefore

$$u(x) \leqslant C_1 e^{-\delta x_n}. \tag{34}$$

The estimate (34) is valid also for $-u(x)$. So we have

$$|u(x)| \leqslant C_1 e^{-\delta x_n} \text{ in } S(0, \infty) .$$

This completes the proof of the theorem.

Theorem 3. *Assume that $K < 0$ and $u(x)$ is a solution of the problem (1),(2) that does not change sign in $S(0, \infty)$.*
   Then if $u(x) > 0$ in $S(0, \infty)$, we have

$$u(x) = M_p' x^{\frac{1}{1-p}} (1 + o(1)) \text{ as } x_n \to \infty ,$$

where

$$M_p' = \left(\frac{|K|}{(p-1)a_0} \right)^{\frac{1}{p-1}} .$$

Proof. Consider the function

$$u_\epsilon(x) = M_p'(1 + \epsilon) x_n^{\frac{1}{1-p}} , \epsilon = const > 0 .$$

We have

$$\sum_{i,j=1}^{n} \frac{\partial}{\partial x_i} \left(a_{ij}(x') \frac{\partial u_\epsilon}{\partial x_j} \right) + \sum_{i=1}^{n-1} a_i(x') \frac{\partial u_\epsilon}{\partial x_i} + K \frac{\partial u_\epsilon}{\partial x_n} - a_0 |u_\epsilon|^{p-1} u_\epsilon$$

$$= M_p'(1 + \epsilon) \frac{p}{(1-p)^2} x_n^{\frac{2p-1}{1-p}} + K M_p'(1 + \epsilon) \frac{p}{1-p} x_n^{\frac{p}{1-p}}$$

$$- a_0 (M_p')^p (1 + \epsilon)^p x_n^{\frac{p}{1-p}} < 0$$

in $S(m, \infty)$, if m is sufficiently large. Therefore, for $V = u_\epsilon(x_n + \tau) - u(x)$ we get

$$\sum_{i,j=1}^{n} \frac{\partial}{\partial x_i} \left(a_{ij}(x') \frac{\partial V}{\partial x_j} \right) + \sum_{i=1}^{n-1} a_i(x') \frac{\partial V}{\partial x_i} + K \frac{\partial V}{\partial x_n} + Q(x)V < 0 , \quad (35)$$

where $Q(x) \leqslant 0$ for $x_n \geqslant m$.
   Let $m_\epsilon = \min_{x_n = m} u_\epsilon(x)$ and X be so large that

$$u(x) \leqslant m_\epsilon \text{ for } x_n = X .$$

We set $\tau = m - X$. From (35) and the boundary condition (2) we have, using the maximum principle, that

$$u(x) \leqslant M_p'(1 + \epsilon)(x_n + m - X)^{\frac{1}{1-p}} \text{ in } S(X, +\infty) . \tag{36}$$

In the same way we consider $u_{-\epsilon} = M_p'(1 - \epsilon)x_n^{\frac{1}{1-p}}$ and we obtain

$$u(x) \geqslant M_p'(1 - \epsilon)(x_n - m + X)^{\frac{1}{1-p}} . \tag{37}$$

From (36), (37) we conclude that

$$u(x) = M_p'(1 + o(1))x_n^{\frac{1}{1-p}} \text{ as } x_n \to \infty .$$

The theorem is proved.

Now let us consider the case $K = 0$. For $K = 0$ and for solutions $u(x)$ of the problem (1),(2), which change sign in $S(0, \infty)$ we have Theorem 1. Now we consider the case, when $K = 0$ and $u(x)$ preserves the sign.

Theorem 4. *If a solution $u(x)$ of the problem (1),(2) is positive in $S(0, \infty)$, $K = 0$, then*

$$u(x) = M_p(x_n + N)^{\frac{2}{1-p}}(1 + \mathcal{O}(\exp\{-hx_n\})) \text{ as } x_n \to \infty \tag{38}$$

where $h = const > 0$ and h does not depend on u, $N = const > 0$ and
$$M_p = \left[\frac{2(1 + p)}{(1 - p)^2 a_0}\right]^{\frac{1}{p-1}} .$$

Proof. Using the maximum principle it is easy to prove that

$$M_p(x_n + \gamma)^{\frac{2}{1-p}} \leqslant u(x) \leqslant M_p x_n^{\frac{2}{1-p}} , \tag{39}$$

where $\gamma = const > 0$. From (39) it follows that the set of nonnegative numbers δ such that

$$u(x) \leqslant M_p(x_n + \delta)^{\frac{2}{1-p}}$$

for $x_n > N_\delta$ is not empty. We denote $\sup \delta$ by δ_0. Consider first the case when $\delta_0 < \infty$. From the definition of δ_0 it follows that

$$u(x) \leqslant M_p(x_n + \delta_0 - \epsilon)^{\frac{2}{1-p}} \tag{40}$$

for any $\epsilon = const > 0$ and for $x_n > X_\epsilon$. In addition,

$$u(x^\epsilon) \geqslant M_p(x_n^\epsilon + \delta_0 + \epsilon)^{\frac{2}{1-p}} , x_n^\epsilon \to \infty . \tag{41}$$

From (41) and the maximum principle we get that for any $x_n > C(\epsilon)$ there exists $x' \in \omega$ such that

$$M_p(x_n + \delta_0 + \epsilon)^{\frac{2}{1-p}} \leqslant u(x', x_n) \ .$$

From (40) it follows that

$$u(x) - M_p(x_n + \delta_0)^{\frac{2}{1-p}}$$
$$\leqslant M_p(x_n + \delta_0 - \epsilon)^{\frac{2}{1-p}} - M_p(x_n + \delta_0)^{\frac{2}{1-p}}$$
$$\leqslant C_1 \epsilon x_n^{(1+p)/(1-p)} \ ,$$

where

$$C_1 = const, x \in S(X_\epsilon, \infty).$$

This means that

$$\left(u(x) - M_p(x_n + \delta_0)^{\frac{2}{1-p}}\right)_+ = o\left(x_n^{(1+p)/(1-p)}\right) \ , \qquad (42)$$

as $x_n \to \infty$, where $f(x)_+ = f(x)$, if $f(x) \geqslant 0$, and $f(x)_+ = 0$, if $f(x) < 0$.

Consider the function

$$v(x) = M_p^{-1}(x_n + \delta_0)^{\frac{2}{(p-1)}} u(x) \ .$$

The function v satisfies the equation

$$\sum_{i,j=1}^n \frac{\partial}{\partial x_i}\left(a_{ij}(x')\frac{\partial v}{\partial x_j}\right) + \sum_{i=1}^{n-1} a_i(x')\frac{\partial v}{\partial x_i}$$
$$+4[(1-p)(x_n+\delta_0)]^{-1}\frac{\partial v}{\partial x_n} + 2(1+p)[(1-p)(x_n+\delta_0)]^{-2}v$$
$$-M_p^{p-1}(x_n+\delta_0)^{-2}|v|^{p-1}v = 0. \qquad (43)$$

From (42) and (39) it follows that

$$(v-1)_+ = o(x_n)^{-1} \text{ as } x_n \to \infty \ ,$$
$$|v(x) - 1| \leqslant Cx_n^{-1} \ , \quad C = const \ . \qquad (44)$$

We set

$$w = v - 1 \ , x_n + \delta_0 = t_n \ , (x_1, \ldots, x_{n-1}) = (t_1, \ldots, t_{n-1}) = \hat{t} \ .$$

The function w satisfies the equation

$$L_2(w) \equiv \sum_{i,j=1}^{n} \frac{\partial}{\partial t_i} \left(a_{ij}(\hat{t}) \frac{\partial w}{\partial t_j} \right) + \sum_{i=1}^{n-1} a_i(\hat{t}) \frac{\partial w}{\partial t_i}$$

$$+ 4((1-p)t_n)^{-1} \frac{\partial w}{\partial t_n} - Q(t)w$$

$$= 0, \tag{45}$$

where

$$Q(t) = \frac{2(1+p)}{(1-p)^2 t_n^2} \left[(w+1)[|w+1|^{p-1} - 1]] \right] w^{-1}$$

$$= \frac{g(t)}{t_n^2} \tag{46}$$

and $g(x)$ is a bounded function, since $|w(x)| \leqslant C x_n^{-1}$. Now we shall use the Harnack inequality in the following form (see [11], [7]). Let $u(x)$ be a weak solution of the equation

$$l(u) \equiv \sum_{i,j=1}^{n} \frac{\partial}{\partial x_i} \left(b_{ij}(x) \frac{\partial u}{\partial x_j} \right) + \sum_{i=1}^{n} b_i(x) \frac{\partial u}{\partial x_i} + b_0(x)u$$

$$= 0$$

in $S(T-2, T+2)$, $T = const > 0$, $u(x) > 0$ in $\overline{S}(T-2, T+2)$,

$$\sum_{i,j=1}^{n} b_{ij}(x)\xi_i\xi_j \geqslant m|\xi|^2, \quad u \in H^1(S(T-2, T+2)),$$

$$m = const > 0, \quad \xi \in \mathbf{R}^n, \quad x \in S(T-2, T+2)),$$

$$\sum_{i,j=1}^{n} |b_{ij}(x)| + \sum_{i=1}^{n} |b_i(x)| + |b_0(x)| \leqslant M, M = const,$$

$|b_0(x)| \leqslant \gamma$, where γ is a sufficiently small constant,

$$\sum_{i,j=1}^{n} b_{ij} \frac{\partial u}{\partial x_j} \nu_i \equiv \frac{\partial u}{\partial \mu} = 0 \text{ on } \sigma(T-2, T+2).$$

Then

$$\max_{S(T-1,T+1)} u(x) \leqslant M_1 \min_{S(T-2,T+2)} u(x), \tag{47}$$

where the constant M_1 depends on m, M, n only.

Consider now the solution w of the equation (45) in $S(T-2, T+2)$, T sufficiently large, and the function $Z(t) = -w(t) + \frac{\alpha}{N}$, where α is a fixed constant, $N \to \infty$. We have $w_+(\hat{t}, N) = o(N^{-1})$, $t \in S(N-2, N+2)$.

Therefore, for sufficiently large N,

$$Z(T) > \frac{\alpha}{2N}, \quad x \in S(N-2, N+2).$$

We shall prove that there exists a point $t_N \in S(N-1, N+1)$ such that

$$Z(t_N) \leqslant \frac{2\alpha}{N} . \tag{48}$$

Indeed, from (41) and the definition of v we have

$$v(x'_\epsilon, x_n) \geqslant M_p^{-1}(x_n + \delta_0)^{\frac{2}{1-p}}(x_n + \delta_0 + \epsilon)^{\frac{2}{(p-1)}} M_p$$

for any $\epsilon > 0$ and some x'_ϵ. Therefore,

$$
\begin{aligned}
w(x'_\epsilon, x_n) &= v(x'_\epsilon, x_n) - 1 \\
&\geqslant (1 + \epsilon(x_n + \delta)^{-1})^{\frac{2}{p-1}} - 1 \\
&\geqslant C_\epsilon(x_n + \delta_0)^{-1} .
\end{aligned}
$$

From this inequality it follows that for sufficiently large N the inequality (48) is valid, if ϵ is sufficiently small.

For the function $Z(t)$ we have the equation

$$
\begin{aligned}
L(Z) &\equiv \sum_{i,j=1}^{n} \frac{\partial}{\partial t_i}\left(a_{ij}(\hat{t})\frac{\partial Z}{\partial t_j}\right) + \sum_{j=1}^{n-1} a_i(\hat{t})\frac{\partial Z}{\partial t_i} \\
&\quad + 4(t_n(1-p))^{-1}\frac{\partial Z}{\partial t_n} - g(t)t_n^{-2}Z \\
&= -\alpha g(t)N^{-1}t_n^{-2} .
\end{aligned}
\tag{49}
$$

Let $Z_0(t)$ be a solution of the following boundary value problem:

$$L(Z_0) = -\alpha g(t)N^{-1}t_n^{-1} \text{ in } S(N-2, N+2) ,$$
$$\frac{\partial Z_0}{\partial \nu} = 0 \text{ on } \sigma(N-2, n+2) ,$$

$Z_0(\hat{t}, t_n) = 0$ for $t_n = N-2$ and for $t_n = N+2$. It is easy to prove, using the maximum principle, that

$$|Z_0(t)| \leqslant \alpha C N^{-3} G \text{ for } t \in S(N-2, N+2), G = \max|g(x)|. \tag{50}$$

Let $Y(t) = Z(t) - Z_0(t)$. For $Y(t)$ we have $L(Y) = 0$ and $Y > 0$ in $S(N-2, N+2)$, $Y(t_N) < \frac{4\alpha}{N}$, $t_N \in S(N-2, N+2)$. Using the Harnack inequality for sufficiently large N, we get

$$Y(t) \leqslant C_1 \alpha N^{-1}, \quad t \in S(N-1, N+1) . \tag{51}$$

Since α is an arbitrary constant, the inequality (51) and (50), (44) imply that

$$Z(t) = o(t_n^{-1}) \text{ as } t_n \to \infty . \tag{52}$$

Therefore,

$$w(t) = o(t_n^{-1}) \text{ as } t_n \to \infty .$$

Let $\theta(t)$ be a function in $\mathbf{R}^1$ such that $\theta(t) = 0$ for $t_n \leqslant T$, $\theta(t) = 1$ for $t \geqslant T+1$ and $\theta \in C^\infty(\mathbf{R}^1)$. The function $w_1(t) = \theta(t_n)w$ satisfies the equation

$$
\begin{aligned}
L_1(w_1) &\equiv \sum_{i,j=1}^{n} \frac{\partial}{\partial t_i}\left(a_{ij}(t)\frac{\partial w_1}{\partial t_j}\right) + \sum_{i=1}^{n-1} a_i(t)\frac{\partial w_1}{\partial t_i} \\
&\quad + a_n(t_n)\frac{\partial w_1}{\partial t_n} + a_0(t)w_1 \\
&= F_0(x) + \sum_{i,j=1}^{n} \frac{\partial F_j}{\partial x_i} \quad \text{in } S(-\infty, +\infty)
\end{aligned} \tag{53}
$$

and the boundary condition $\dfrac{\partial w_1}{\partial \mu} = 0$ on $\sigma(-\infty, +\infty)$, where F_i is a function with a compact support, $i = 0, 1, \ldots, n$, $F_i \in L_2(S(-\infty, +\infty))$,

$$a_n(t_n) = 4((1-p)t_n)^{-1}$$

for $t_n > T+1$, $a_n = 0$ for $t_n < T$, $a_0(t) = -g(t)t_n^{-2}$ for $t_n > T+1$, $a_0(t) = 0$ for $t_n < T$, the function $g(t)$ is defined by (46). From (46) and (52) it follows that for $t_n \geqslant T+1$

$$a_0(t) = \frac{2(1+p)}{t_n^2(1-p)}(1 + \mathcal{O}(t_n^{-1})) \tag{54}$$

and, therefore, $a_0(t) < 0$, if T is sufficiently large. By Proposition 1, if ϵ_1 is a number such that on the line $\text{Im}\,\lambda = \epsilon_1$ there is no eigenvalue of problem (15),(16) corresponding to the operator L_1 with the boundary condition $\dfrac{\partial w_1}{\partial \mu} = 0$ on $\delta\omega$, then there exists in $S(-\infty, +\infty)$ a solution w_2

of equation (53) with the boundary condition $\dfrac{\partial w_2}{\partial \mu} = 0$ on $\sigma(-\infty, \infty)$ such that $I_{\epsilon_1}(w_2) < \infty$.

The function $w_3(t) = \theta_1(t_n)w_2$, where $\theta_1(t_n) = 1$ for $t_n < T - 1$, $\theta_1(t_n) = 0$ for $t_n > T$, $\theta_1 \in C^\infty(\mathbf{R}^1)$, satisfies an equation of the form (53) with $a_0 \equiv 0$, $a_n \equiv 0$ in $S(-\infty, \infty)$. From Proposition 2 it follows that

$$w_3(t) = a + bt_n + \beta(t), \tag{55}$$

where a, $b = const$, $I_{-\epsilon_1}(\beta) < \infty$. We note that $w_3 = w_2$ for $t_n < T - 1$. From (55) it follows that $w_2 = a + bt_n + \beta(t)$ and $|\beta(t)| \leqslant C \exp\{\epsilon_1 t_n\}$ for $t_n \leqslant T - 1$, where ϵ_1 is the constant defined above. Consider the function $W(t) = w_2(t) - w_1(t)$. Since $I_{\epsilon_1}(w_2) < \infty$, we have $|w_2(t)|^2 \leqslant C_1 \exp\{-2\epsilon_1 t_n\}$ for $t_n > T$. In addition, $W(t) \to 0$ as $t_n \to \infty$, because $w_1 \to 0$ as $t_n \to \infty$. Let us prove that $a = 0$, $b = 0$ in (55). Suppose that $a \neq 0$ or $b \neq 0$. Then we get that $W > 0$ for $t_n < T_1$, $T_1 = const$, changing the sign of W, if necessary. According to the maximum principle $W(t) > 0$ in $S(-\infty, \infty)$, since $w_2(t) \to 0$, $w_1(t) \to 0$ as $t_n \to \infty$. The function W satisfies the equation $L_1(W) = 0$ in $S(-\infty, \infty)$, $\dfrac{\partial W}{\partial \mu} = 0$ on $\sigma(-\infty, \infty)$. For $t_n > T_1$, $\ae > 0$, and T_1 and $\ae$ sufficiently large, the function $V(t) = t_n^{-1}(1 + \dfrac{\ae^2}{t_n^2})$ satisfies the inequality $L_1(V) \geqslant 0$. This is easy to verify, taking into account that

$$\begin{aligned} a_n(t) &= -4(p-1)^{-1}t_n^{-1}, \\ a_0(t) &= \frac{2(1+p)}{t_n^2(1-p)}\left(1 + \mathcal{O}\left(\frac{1}{t_n}\right)\right), \\ a_{nn} &= 1. \end{aligned}$$

Let us define γ in such a way that $W \geqslant \gamma V$ for $t_n = T_1$. Since $W \to 0$ and $V \to 0$ as $t_n \to \infty$, by the maximum principle we have $W \geqslant \gamma V$ in $S(T_1, \infty)$. Therefore, $W(t) \geqslant \gamma t_n^{-1}$, and we have a contradiction, because $W(t) = o(t_n^{-1})$ as $t_n \to \infty$, since $W = w_2 - w_1$, $|w_2(x)| \leqslant C_2 \exp\{-\epsilon_1 t_n\}$, $w_1 = w$ for $t_n > T$ and $|w(t)| = o(t_n^{-1})$ as $t_n \to \infty$. Therefore $W(t) \to 0$ as $t_n \to \infty$ and $t_n \to -\infty$. By the maximum principle $W \equiv 0$ in $S(-\infty, \infty)$. We have $w_1 = w = w_2$ for $t_n > T + 1$. It was proved that $|w_2(t)| \leqslant C_3 \exp\{-\epsilon_1 t_n\}$. Thus,

$$v(x) = 1 + \mathcal{O}(\exp\{-\epsilon_1 x_n\}),$$

and

$$u(x) = M_p(x_n + \delta_0)^{2/(1-p)}(1 + \mathcal{O}(\exp\{-\epsilon_1 x_n\})) \ .$$

The theorem is thus proved for the case $\delta_0 < \infty$. Consider the case $\delta_0 = \infty$. Since $u(x) > 0$ in $S(0, \infty)$, according to (39) we get

$$u(x) \geqslant M_p(x_n + \gamma_1)^{2/(1-p)} .$$

Therefore, for any $\delta > 0$ which satisfies the inequality

$$u(x) \leqslant M_p(x_n + \delta)^{2/(1-p)} .$$

we have $\delta < \gamma_1$, which contradicts the assumption that $\delta_0 = \infty$. The theorem is proved.

Another proof of Theorem 1 for $K = 0$ is given in [10].

Now we consider a more general class of nonlinear elliptic equations in $S(0, \infty)$:

$$\begin{aligned} L(U) - a_0|U|^{p-1}U \\ \equiv \sum_{i,j=1}^{n} \frac{\partial}{\partial x_i}\left(a_{ij}(x')\frac{\partial U}{\partial x_j}\right) + \sum_{i,j=1}^{n} a_j(x')\frac{\partial U}{\partial x_j} - a_0|U|^{p-1}U = 0, \end{aligned} \tag{56}$$

where

$$m_1|\xi|^2 \leqslant \sum_{i,j=1}^{n} a_{ij}(x')\xi_i\xi_j \leqslant m_2|\xi|^2, \quad m_1, m_2 = const > 0,$$

$a_{ij}(x') = a_{ji}(x')$, $a_{nn} \equiv 1$, $a_{in} = 0$ for $i < n$, $a_0 = const > 0$, $p = const > 1$. Here we assume that $a_n(x')$ may change the sign in ω.

We study an asymptotic behavior as $x_n \to \infty$ of solutions $u(x)$ of the equation (56) in $S(0, \infty)$ with the boundary condition

$$\frac{\partial U}{\partial \mu} \equiv \sum_{i,j=1}^{n} a_{ij}(x')\frac{\partial U}{\partial x_j}\nu_i = 0 \text{ on } \sigma(0, \infty), \tag{57}$$

where $\nu = (\nu_1, \dots, \nu_n)$ is an exterior unit normal vector to $\sigma(0, \infty)$.

Consider the auxiliary problem

$$\mathcal{L}(U) \equiv \sum_{i,j=1}^{n-1} \frac{\partial}{\partial x_i}\left(a_{ij}(x')\frac{\partial U}{\partial x_j}\right) + \sum_{j=1}^{n-1} a_j(x')\frac{\partial U}{\partial x_j} = f(x') \text{ in } \omega, \tag{58}$$

$$\frac{\partial U}{\partial \mu} \equiv \sum_{i,j=1}^{n-1} a_{ij}(x')\frac{\partial U}{\partial x_j}\nu_i = \varphi(x') \text{ on } \partial\omega, \tag{59}$$

where $f(x') \in L^2(\omega)$, $\varphi(x') \in L^2(\partial\omega)$. The problem (58),(59) has a solution if and only if

$$\int_\omega f(x')v_0(x')\mathrm{d}x' = \int_{\partial\omega} \varphi(x')v_0(x')\mathrm{d}s, \qquad (60)$$

where $v_0(x')$ is a solution of the problem

$$\mathcal{L}^*(v) \equiv \sum_{i,j=1}^{n-1} \frac{\partial}{\partial x_i}\left(a_{ij}(x')\frac{\partial v}{\partial x_j}\right) - \sum_{j=1}^{n-1}\frac{\partial}{\partial x_j}(a_j(x')v) = 0 \text{ in } \omega, \qquad (61)$$

$$\frac{\partial v}{\partial \mu} - \sum_{j=1}^{n-1} a_j(x')\nu_j v = 0 \text{ on } \partial\omega. \qquad (62)$$

The problem (61),(62) is the adjoint problem to the problem (58),(59) and the condition (60) is necessary and sufficient by virtue of the Fredholm theory. The function $v_0(x')$ does not change the sign in ω. We suppose that $v_0(x') > 0$ in ω. If $a_j(x') \equiv 0$ in ω for $j = 1, \ldots, n-1$, then the problem (58),(59) is self-adjoint and $v_0(x') \equiv 1$.

For simplicity we assume that coefficients $a_{ij}(x')$, $a_j(x')$ are sufficiently smooth in $\overline{\omega}$ and we consider solutions $U(x)$ of the problem (56),(57) which belong to the class $C^2(S(0,\infty)) \cap C^1(\overline{S(0,\infty)})$.

We shall use the following comparison theorem.

Theorem 5. *Assume that* $V(x) > 0$ *in* $S(0,\infty)$ *and such that*

$$L(V) - a_0 V^p \equiv \sum_{i,j=1}^{n} \frac{\partial}{\partial x_i}\left(a_{ij}(x')\frac{\partial V}{\partial x_j}\right) + \sum_{j=1}^{n} a_j(x')\frac{\partial V}{\partial x_j} - a_0 V^p \leqslant 0 \quad (63)$$

in $S(0,\infty)$,

$$\frac{\partial V}{\partial \mu} = 0 \text{ on } \sigma(0,\infty), \quad V(x) \geqslant u(x) \text{ for } x_n = 0, \qquad (64)$$

where $u(x)$ *is a solution of problem (56),(57). Then*

$$V(x) \geqslant u(x) \text{ for } x \in S(0,\infty).$$

If $V(x) > 0$ *in* $S(0,\infty)$,

$$L(V) - a_0 V^p \geqslant 0 \text{ in } S(0,\infty), \qquad (65)$$

$\frac{\partial V}{\partial \mu} = 0$ *for* $x \in \sigma(0,\infty)$, $V(x) \to 0$ *as* $x_n \to \infty$, $V(x) \leqslant u(x)$ *for* $x_n = 0$, *then*

$$u(x) \geqslant V(x) \text{ for } x \in S(0,\infty). \qquad (66)$$

The function $V(x)$ satisfying conditions (63),(64) is called a supersolution or an upper function. The function $V(x)$ satisfying conditions (65),(66) is called a subsolution or a lower function.

Theorem 5 can be easily proved using the maximum principle for linear equations.

Theorem 6. *Let $U(x)$ be a solution of problem (56),(57) and let*

$$\int_\omega a_n(x')v_0(x')\mathrm{d}x' < 0.$$

Then for any $\epsilon > 0$ there exist constants T and τ such that

$$|U(x)| \leqslant (K_p + \epsilon)(x_n - T)^{\frac{1}{1-p}} \text{ for } x \in S(\tau + T, \infty),$$

where

$$K_p = \left[\frac{p \int\limits_\omega a_n(x')v_0(x')\mathrm{d}x'}{(1-p)a_0 \int\limits_\omega v_0(x')\mathrm{d}x'} \right]^{\frac{1}{1-p}}.$$

If

$$\int_\omega a_n(x')v_0(x')\mathrm{d}x' \geqslant 0,$$

then for any $\epsilon > 0$ there exist constants T and τ such that

$$|U(x)| \leqslant \epsilon(x_n - T)^{\frac{1}{1-p}} \text{ for } x \in S(T + \tau, \infty).$$

Proof. Consider the function

$$V(x) = cx_n^{\frac{1}{1-p}} + x_n^{\frac{p}{1-p}} \phi(x').$$

Let us choose the constant c and the function $\phi(x')$ in such a way that the function $V(x)$ is a supersolution for problem (56),(57). It is easy to see that

$$L(V) - a_0 V^p = x_n^{\frac{p}{1-p}} \left\{ \frac{c}{1-p} a_n(x') - a_0 c^p + \mathcal{L}(\phi) \right\} (1 + o(1))$$

as $x_n \to \infty$. We look for the function $\phi(x')$ as a solution of the problem

$$\mathcal{L}(\phi) + \frac{c}{1-p} a_n(x') - a_0 c^p = -\epsilon_1^2, \quad \epsilon_1 = const > 0, x' \in \omega, \qquad (67)$$

$$\frac{\partial \phi}{\partial \mu} = 0 \quad \text{on } \partial \omega. \tag{68}$$

The problem (67),(68) has a solution, if

$$\int_\omega \frac{c}{1-p} a_n(x') v_0(x') \mathrm{d}x' - a_0 c^p \int_\omega v_0(x') \mathrm{d}x' = -\epsilon_1^2 \int_\omega v_0(x') \mathrm{d}x'. \tag{69}$$

Assume that

$$\int_\omega a_n(x') v_0(x') \mathrm{d}x' < 0, \tag{70}$$

and set $c = K_p + \epsilon$. For such c we define from (69) the constant ϵ_1. If

$$\int_\omega a_n(x') v_0(x') \mathrm{d}x' \geqslant 0, \tag{71}$$

then, taking $c = \epsilon$, we can find from (69) the constant ϵ_1. The function $V(x)$, constructed above, satisfies the inequality

$$L(V) - a_0 V^p < 0 \quad \text{in } S(\tau, \infty),$$

if τ is sufficiently large, and $\frac{\partial V}{\partial \mu} = 0$ on $\sigma(\tau, \infty)$. If T is sufficiently large, then $V(x', \tau) > u(x', T + \tau)$. According to the comparison theorem

$$V(x', x_n + \tau) \geqslant u(x', T + \tau + x_n) \quad \text{in } S(0, \infty). \tag{72}$$

It follows from (72) that

$$U(x) \leqslant (K_p + \epsilon)(x_n - T)^{\frac{1}{1-p}} \quad \text{in } S(\tau + T, \infty),$$

if (70) is satisfied, and

$$U(x) \leqslant \epsilon(x_n - T)^{\frac{1}{1-p}} \quad \text{in } S(\tau + T, \infty),$$

if (71) is valid.

In a similar way we can prove the same inequalities for $-U(x)$ and the assertion of Theorem 6.

Theorem 7. *If $\int_\omega a_n(x') v_0(x') \mathrm{d}x' < 0$ and $U(x) > 0$ is a solution of the problem (56),(57), then in $S(0, \infty)$*

$$U(x) = K_p(1 + o(1))(x_n - T)^{\frac{1}{1-p}} \quad \text{as } x_n \to \infty.$$

Proof. We construct the lower function in the form

$$V(x) = cx_n^{\frac{1}{1-p}} + x_n^{\frac{p}{1-p}}\phi(x'),$$

where $c = K_p - \epsilon$, $\epsilon = const > 0$,

$$\mathcal{L}(\phi) + c\lambda a_n(x') - a_0 c^p = -\epsilon_1^2, \quad x' \in \omega, \lambda = \frac{1}{1-p}, \tag{73}$$

$$\frac{\partial \phi}{\partial \mu} = 0 \quad \text{on } \partial\omega. \tag{74}$$

The problem (73),(74) has a solution, if

$$c\lambda \int_\omega a_n(x')v_0(x')\mathrm{d}x' - a_0 c^p \int_\omega v_0(x')\mathrm{d}x' = -\epsilon_1^2 \int_\omega v_0(x')\mathrm{d}x'.$$

This equality defines ϵ_1 for small ϵ. It is easy to see that the function $V(x)$ satisfies the following conditions:

$$L(V) - a_0 V^p > 0, \quad V > 0 \text{ in } S(T,\infty), \quad \frac{\partial V}{\partial \mu} = 0 \quad \text{on } \sigma(T,\infty),$$

$V(x) \to 0$ as $x_n \to \infty$. If T is sufficiently large, then also

$$U(x',0) > V(x',T).$$

According to Theorem 1

$$U(x) \geqslant (K_p - \epsilon)(x_n + T)^{\frac{1}{1-p}} \quad \text{in } S(T,\infty). \tag{75}$$

Theorem 7 follows from (75) and Theorem 6.

Theorem 8. *If $\int_\omega a_n(x')v_0(x')\mathrm{d}x' > 0$, U is a solution of the problem (56),(57), then*

$$|U(x)| \leqslant C_1 e^{-\alpha x_n},$$

where C_1, α are positive constants and α does not depend on u.

Proof. Let us look for an upper function of the form

$$V(x) = e^{-\epsilon x_n}(1 + \epsilon\phi(x')), \quad \epsilon = const > 0.$$

For $V(x)$ to be an upper function, it is necessary that $V > 0$ and

$$
\begin{aligned}
L(V) - a_0 V^p &= (\epsilon^2 - \epsilon a_n(x') + \epsilon \mathcal{L}(\phi)) e^{-\epsilon x_n} - a_0 e^{-\epsilon p x_n} (1 + \epsilon \phi(x'))^p \\
&\leqslant 0 \text{ in } S(0, \infty).
\end{aligned}
$$

Let $\phi(x')$ be such that

$$
\mathcal{L}(\phi) = a_n(x') + \beta \quad \text{in } \omega,
$$

$$
\frac{\partial \phi}{\partial \mu} = 0 \quad \text{on } \partial \omega.
$$

This problem has a solution, if

$$
\int_\omega a_n(x') v_0(x') \mathrm{d}x' + \beta \int_\omega v_0(x) \mathrm{d}x' = 0.
$$

Defining β, which turns out to be negative, by the above relation, we obtain $V(x) > 0$, $L(V) - a_0 V^p \leqslant 0$ in $S(T, \infty)$, if $\epsilon > 0$ is sufficiently small and $T(\epsilon)$ is sufficiently large. If τ is sufficiently large then

$$
V(x', x_n + T) > U(x', x_n + T + \tau) \quad \text{for } x_n = 0
$$

It follows by the comparison theorem that

$$
U(x', x_n) \leqslant V(x', x_n - T) \quad \text{for } x_n > T + \tau
$$

In an analogous manner it can be established that $-u(x', x_n) \leqslant V(x', x_n - T)$ for $x_n > T + \tau$. The theorem is proved.

Let us pass to the case

$$
\int_\omega a_n(x') v_0(x') \mathrm{d}x' = 0. \tag{76}
$$

This case is more delicate.

Theorem 9. *If the condition (76) is satisfied, then for solutions $U(x)$ of the problem (56),(57) we have*

$$
|U(x)| = O\left(x_n^{\frac{2}{1-p}}\right) \quad \text{as } x_n \to \infty.
$$

Proof. Let us look for an upper function of the form

$$V(x) = cx_n^\lambda + x_n^{\lambda-1}\phi_1(x') + x_n^{\lambda-2}\phi_2(x'), \quad \lambda = \frac{2}{1-p}.$$

The conditions

$$L(V) - V^p \leqslant 0, V > 0 \text{ in } S(T,\infty), \ \frac{\partial V}{\partial \mu} = 0 \text{ on } \sigma(T,\infty)$$

will be satisfied, if $c > 0$,

$$\lambda c a_n(x') + \mathcal{L}(\phi_1(x')) = 0 \text{ in } \omega, \ \frac{\partial \phi_1}{\partial \mu} = 0 \text{ on } \partial\omega \qquad (77)$$

and

$$-c^p a_0 + \lambda(\lambda-1)c + (\lambda-1)a_n(x')\phi_1(x') + \mathcal{L}(\phi_2(x')) = -\epsilon_1^2 \text{ in } \omega, \quad (78)$$

$$\epsilon_1 = const > 0, \ \frac{\partial \phi_2}{\partial \mu} = 0 \text{ on } \partial\omega. \qquad (79)$$

The problem (77) is solvable for any c by virtue of the conditions of Theorem 9. The condition for the solvability of the problem (78),(79) is given by the following relation:

$$-c^p a_0 \int_\omega v_0(x')\mathrm{d}x' + (\lambda-1) \int_\omega a_n(x')v_0(x')\phi_1(x')\mathrm{d}x'$$

$$+\lambda(\lambda-1)c \int_\omega v_0(x')\mathrm{d}x' < 0. \qquad (80)$$

Let $\phi_3(x')$ be a solution of the problem

$$a_n(x') + \mathcal{L}(\phi_3) = 0 \text{ in } \omega, \ \frac{\partial \phi_3}{\partial \mu} = 0 \text{ on } \partial\omega.$$

Then $\phi_1(x') = c\lambda\phi_3(x')$. If c is taken to be sufficiently large, then (80) will be satisfied. In that case

$$V(x',T) > U(x',T+\tau)$$

if τ is sufficiently large and by the comparison theorem we have

$$V(x',x_n+T) > U(x',x_n+T+\tau) \text{ for } x_n > 0.$$

Therefore $U \leqslant C_0 x_n^{\frac{2}{1-p}}$. Similarly, it can be shown that $-U \leqslant C_0 x_n^{\frac{2}{1-p}}$. Thus, $U = O(x_n^{\frac{2}{1-p}})$ as $x_n \to \infty$.

Remark. The equation

$$-c^p a_0 \int\limits_\omega v_0(x')\mathrm{d}x' + c(\lambda - 1)\lambda \int\limits_\omega a_n(x')\phi_3(x')v_0(x')\mathrm{d}x'$$

$$+\lambda(\lambda - 1)c \int\limits_\omega v_0(x')\mathrm{d}x' = 0$$

has one positive root $C_1(\omega)$, if

$$\int\limits_\omega a_n(x')\phi_3(x')\mathrm{d}x' + \int\limits_\omega v_0(x')\mathrm{d}x' > 0. \tag{81}$$

If we put $c = C_1(\omega) + \epsilon$ in (80), then under conditions of Theorem 5 we obtain
$$|u(x)| \leqslant (C_1(\omega) + \epsilon)x_n^{\frac{2}{1-p}}(1 + o(x_n)) \quad \text{as } x_n \to \infty.$$

Theorem 10. *If the condition (81) holds,*

$$\int\limits_\omega a_n(x')v_0(x')\mathrm{d}x' = 0,$$

$u(x) > 0$ *is solution of the problem (56),(57), then*

$$u(x) = (C_1(\omega) + \epsilon)x_n^{\frac{2}{1-p}}(1 + o(1)) \quad \text{as } x_n \to \infty. \tag{82}$$

Proof. It follows from the remark after Theorem 9 that

$$u(x) \leqslant (C_1(\omega) + \epsilon)x_n^{\frac{2}{1-p}} \quad \text{in } S(T,\infty), \quad T = const > 0.$$

Let us construct a lower function $V(x)$ of the form

$$V(x) = cx_n^\lambda + x_n^{\lambda-1}\phi_1(x') + x_n^{\lambda-2}\phi_2(x'), \quad \lambda = \frac{2}{1-p}.$$

Choose $\phi_1(x')$ such that the relation (77) are satisfied. The function $\phi_2(x')$ is taken to be a solution of the problem

$$-c^p a_0 + \lambda(\lambda - 1)c + (\lambda - 1)a_n(x')\phi_1(x') + \mathcal{L}(\phi_2) = \epsilon_2^2 \text{ in } \omega, \tag{83}$$

$$\frac{\partial \phi_2}{\partial \mu} = 0 \quad \text{on } \partial \omega, \tag{84}$$

ϵ_2 is a positive constant. Let $c = C_1(\omega) - \epsilon$. Then problem (83),(84) can be solved because

$$-c^p a_0 \int_\omega v_0(x') \mathrm{d}x' + c\lambda(\lambda - 1) \int_\omega v_0(x') \mathrm{d}x'$$

$$+c\lambda(\lambda - 1) \int_\omega a_n(x') \phi_3 v_0(x') \mathrm{d}x' > 0.$$

Here the function $\phi_3(x')$ is a solution of the problem:

$$a_n(x') + \mathcal{L}(\phi_3) = 0 \quad \text{in } \omega, \quad \frac{\partial \phi_3}{\partial \mu} = 0 \quad \text{on } \partial \omega.$$

The function $V(x)$ constructed in this way is a lower function and consequently

$$u(x) \geqslant (C_1(\omega) - \epsilon) x_n^{\frac{2}{1-p}}. \tag{85}$$

The equality (82) follows from (85) and the remark after Theorem 9. Theorem 10 is proved.

We state now several known lemmas without proofs. They are a generalization of Propositions 1 and 2.

Lemma 1. *Suppose that the line* $\operatorname{Im} \lambda = h$ *does not contains any eigenvalue of the eigenvalue problem:*

$$\mathcal{L}(u) - \lambda^2 u + i\lambda a_n(x')u = 0 \quad \text{in } \omega, \quad \frac{\partial u}{\partial \mu} = 0 \quad \text{on } \partial \omega. \tag{86}$$

Let

$$J_h(f_1, f_2) = \int_{S(-\infty, \infty)} e^{2hx_n}(f_1^2(x) + f_2^2(x)) \mathrm{d}x < \infty.$$

Then there exists a unique solution of the equation

$$L(u) = f_1 + \frac{\partial f_2}{\partial x_n} \quad \text{in } S(-\infty, +\infty) \tag{87}$$

such that

$$\frac{\partial u}{\partial \mu} = 0 \quad \text{on } \sigma(-\infty, +\infty) \tag{88}$$

and

$$I_h(u) \equiv \int\limits_{S(-\infty,\infty)} e^{2hx_n}(|u|^2 + |\nabla u|^2)\mathrm{d}x < \infty.$$

In addition

$$I_h(u) \leqslant CJ_h(f_1, f_2), \quad C = const. \tag{89}$$

This lemma is easily proved by using the Fourier transform with respect to x_n.

Lemma 2. *Let h be such that the line* $\mathrm{Im}\,\lambda = h$ *contains no points of the spectrum of problem (86). There exists* $\epsilon = const > 0$ *such that, if*

$$|a_i(x') - a_i^*(x)| \leqslant \epsilon, \ |a_0^*(x)| \leqslant \epsilon, \ J_h(f_1, f_2) < \infty, i = 1, \ldots, n,$$

then the equation

$$L_\epsilon(u) \equiv \sum_{i,j}^n \frac{\partial}{\partial x_i}\left(a_{ij}(x')\frac{\partial u}{\partial x_j}\right) + \sum_{i=1}^n a_i^*(x)\frac{\partial u}{\partial x_i} + a_0^*(x)u = f_1 + \frac{\partial f_2}{\partial x_n} \tag{90}$$

has a unique solution in $S(-\infty,\infty)$ *satisfying the condition (88) and* $I_h(u) < \infty$. *Moreover, inequality (89) holds.*

This lemma follows from Lemma 1 and from the theorem on the invertibility of an operator close to an invertible operator.

Lemma 3. *Let* h_1, h_2 *be such constants that there are no points of the spectrum of the problem (86) on the lines* $\mathrm{Im}\,\lambda = h_1$ *and* $\mathrm{Im}\,\lambda = h_2$. *Further, let*

$$J_{h_1}(f_1, f_2) + J_{h_2}(f_1, f_2) < \infty, \quad I_{h_1}(u) < \infty,$$

where u(x) is a solution of the problem (87),(88). Then

$$\begin{aligned} u(x) = & \sum_{h_1 < \mathrm{Im}\,\lambda_j < h_2} e^{i\lambda_j x_n}\phi_j(x')x_n^{K_j} \\ & + \sum_{\substack{h_1 < \mathrm{Im}\,\lambda_j < h_2 \\ 1 \leqslant K \leqslant K_j}} e^{j\lambda_j x_n}\phi_{jK}(x')x_n^{K_j-K} + u_1(x), \end{aligned}$$

where λ_j *are eigenvalues of the problem (86),* $\phi_j(x')$ *is an eigenfunction and* $\phi_{jK}(x')$ *(K = 1, \ldots, K_j) is an adjoint function of the problem (86), corresponding to* $\lambda = \lambda_j$,

$$I_{h_2}(u_1) \leqslant C[J_{h_1}(f_1, f_2) + J_{h_2}(f_1, f_2)], \quad C = const.$$

Theorem 11. *If $u(x)$, the solution of the problem (56),(57), changes sign in every domain $S(a, \infty)$, $a > 0$, then*

$$|u(x)| \leqslant Ce^{-\alpha x_n} \quad in \quad S(0, \infty),$$

where the constant α does not depend on u.

Proof. Let us remark that the problem (86) always has the eigenvalue $\lambda = 0$ with the corresponding eigenfunction $\psi(x') \equiv 1$. If

$$\int_\omega a_n(x')v_0(x')dx' \neq 0,$$

then there is no adjoint function corresponding to the eigenvalue $\lambda = 0$. If

$$\int_\omega a_n(x')v_0(x')dx' = 0,$$

then there exists an adjoint function ψ_1 which is a solution to the problem

$$\mathcal{L}(\psi_1) + ia_n(x') = 0 \quad in \quad \omega, \qquad \frac{\partial \psi_1}{\partial \mu} = 0 \quad on \quad \partial\omega.$$

We shall show now that the eigenvalue $\lambda = 0$ cannot have another adjoint function. If ψ_2 is another adjoint function with $\lambda = 0$, then

$$\mathcal{L}_2(\psi_2) + ia_n(x')\psi_1 - 1 = 0 \quad in \quad \omega, \qquad \frac{\partial \psi_2}{\partial \mu} = 0 \quad on \quad \partial\omega.$$

In this case, the function

$$W(x) = \frac{1}{2}x_n^2 - ix_n\psi_1(x') - \psi_2(x')$$

can be shown to be a solution to the problem

$$L(W) = 0 \quad in \quad S(-\infty, \infty), \qquad \frac{\partial W}{\partial \mu} = 0 \quad on \quad \sigma(-\infty, \infty),$$

which contradicts the minimum principle.

Note that $\psi_1(x')$ is a purely imaginary function. Thus, the eigenvalue $\lambda = 0$ of the problem (86) has only one adjoint function, corresponding to $\lambda = 0$.

To prove Theorem 11 we may confine ourselves to the case

$$\int_{\omega} a_n(x')v_0(x')dx' \leqslant 0,$$

since in the contrary case, every solution of the problem (56),(57) decreases at an exponential rate by Theorem 8.

Consider the function $v(x) = \theta(x_n)u(x)$, where $u(x)$ is a solution to problem (56),(57), $\theta(x_n) \in C^{\infty}(\mathbf{R}_1)$, $\theta(x_n) = 0$ for $x_n < T$, $\theta(x_n) = 1$ for $x_n > T + 1$, $T = const$. The function $v(x)$ satisfies the equation

$$L(v) - q(x)v = f_1 + \frac{\partial f_2}{\partial x_n} \quad \text{in} \quad S(-\infty, +\infty) \tag{91}$$

where f_1 and f_2 are bounded functions with compact support in x_n. The function $v(x)$ satisfies the boundary condition

$$\frac{\partial v}{\partial \mu} = 0 \quad \text{on} \quad \sigma(-\infty, \infty). \tag{92}$$

The coefficient $q(x)$ of the equation (91) equals to $\theta(x)|u|^{p-1}$ and, if T is sufficiently large, we have $|q(x)| < \epsilon$, since $u(x) \to 0$ as $x_n \to \infty$.

From Lemma 2 it follows that the problem (91),(92) has a solution $W(x)$ such that

$$\int_{S(-\infty, \infty)} e^{2h_1 x_n}(|W|^2 + |\nabla W|^2)dx < \infty, \quad h_1 = const > 0. \tag{93}$$

From (93) and the De Giorgi estimate [9],[7] it follows that

$$W(x) = O(e^{-h_1 x_n}) \quad \text{as} \quad x_n \to \infty.$$

In fact, the number h_1 in (93) can be taken to be any number such that the strip $0 < \text{Im}\,\lambda \leqslant h_1$ does not contain any point of the spectrum of the problem (86). Consider the difference $w(x) = W(x) - v(x)$. This function is a solution of the equation

$$L(w) - q(x)w = 0 \quad \text{in} \quad S(-\infty, \infty), \quad \frac{\partial w}{\partial \mu} = 0 \quad \text{on} \quad \sigma(-\infty, \infty).$$

The function $q(x)$ may be regarded as a predetermined zero for $x_n < T$. From Lemma 3 we then have

$$w(x) = A(x_n + \psi_1(x')) + B + w_0(x), \quad A, B = const, \tag{94}$$

where

$$\int\limits_{S(-\infty,+\infty)} e^{2h_2 x_n}(w_0^2 + |\nabla w_0|^2)\mathrm{d}x < \infty, \qquad (95)$$

h_2 is any number such that in the strip $h_2 \leqslant \operatorname{Im}\lambda < 0$ there are no points of the spectrum of the problem (86). It can be proved in the same way as in Theorem 1 that for sufficiently small h_1 and $|h_2|$ only $\lambda = 0$ is the eigenvalue of the problem (86) in the strip $h_2 \leqslant \operatorname{Im}\lambda \leqslant h_1$. From (95) and the De Giorgi estimates for solutions of elliptic equations we obtain

$$|w_0(x)| \leqslant Ce^{-h_2 x_n}, \quad C = const.$$

In the representation (94) $\psi_1(x')$ is the adjoint function of the eigenvalue problem (86) corresponding to the eigenvalue $\lambda = 0$. In the case when such adjoint functions do not exist, we have $A = 0$ in (94). We now show that in (94) we always have $A = 0$. Suppose that $A \neq 0$. First, let $A < 0$. In this case $w(x) > 0$ for $x_n \leqslant -T_1$, $T_1 = const > 0$. From the maximum principle it follows that $w(x) > 0$ in $S(-\infty,\infty)$.

To estimate $w(x)$ from below, let us construct a lower function of the form

$$V(x) = x_n^\gamma + x_n^{\gamma-1}\varphi(x'), \quad \gamma = const < 0.$$

Choose λ, $\varphi(x')$ such that

$$L(V) - qV \geqslant 0 \quad \text{in} \quad S(T,\infty), \qquad \frac{\partial V}{\partial \mu} = 0 \quad \text{on} \quad \sigma(T,\infty). \qquad (96)$$

Recall that $|q(x)| \leqslant C_1 x_n^{-1}$ in view of Theorem 2 and 5. The relations (96) will be satisfied if

$$\gamma a_n(x') + \mathcal{L}(\varphi(x')) - C_1 \geqslant \delta_1^2 \quad \text{for} \quad x' \in \omega, \qquad (97)$$

$$\frac{\partial \varphi}{\partial \mu} = 0 \quad \text{on} \quad \partial\omega. \qquad (98)$$

The solvability condition for the problem (97),(98) is

$$\gamma \int\limits_\omega a_n(x')v_0(x')\mathrm{d}x' - C_1 \int\limits_\omega v_0(x')\mathrm{d}x' > 0.$$

So, if $\int\limits_\omega a_n(x')v_0(x')\mathrm{d}x' < 0$, then choosing γ to be negative and sufficiently large in modulus, we see that conditions (97),(98) are satisfied and thus the lower function is constructed. Let

$$\int_\omega a_n(x')v_0(x')\mathrm{d}x' = 0. \tag{99}$$

In this case we seek a lower function of the form

$$V(x) = x_n^\gamma + x_n^{\gamma-1}\varphi_1(x') + x_n^{\gamma-2}\varphi_2(x').$$

For the condition (96) to be satisfied, it is sufficient to choose γ, φ_1, φ_2 such that

$$\gamma a_n(x') + \mathcal{L}(\varphi_1(x')) = 0 \quad \text{in} \quad \omega, \tag{100}$$

$$\gamma(\gamma-1) + (\gamma-1)\varphi_1 a_n + \mathcal{L}(\varphi_2) - C_1 = \delta^2 > 0 \quad \text{in} \quad \omega, \tag{101}$$

$$\frac{\partial\varphi_1}{\partial\mu} = 0, \quad \frac{\partial\varphi_2}{\partial\mu} = 0 \quad \text{on} \quad \partial\omega. \tag{102}$$

The equation (100) with the condition (102) can be solved, since the assumption (99) is valid.

In order that the equation (101) with the condition (102) be solvable, it is sufficient to take γ negative and sufficiently large in modulus.

Thus, in all cases, using the lower function $V(x)$, we obtain the inequality

$$w(x) \geqslant C_2 x_n^\gamma \quad \text{in} \quad S(\tau,\infty),$$

where $\gamma < 0$ and τ is sufficiently large. Hence

$$W(x) \geqslant C_2 x_n^\gamma + v(x). \tag{103}$$

Since the relation $|W(x)| \leqslant Ce^{-h_1 x_n}$, $h_1 = const > 0$, holds the inequality (103) is not possible at the points, where $v(x) = u(x) = 0$. By assumption such an infinite sequence of points which tends to ∞ exists. If $A > 0$ a similar argument also leads to a contradiction. Thus, in (94) $A = 0$. In the same way we prove that $B = 0$. So we have that $w(x) \to 0$ as $x_n \to +\infty$ and $x_n \to -\infty$. From the maximum principle we obtain that $w(x) \equiv 0$ in $S(-\infty,\infty)$. This implies that $v(x) = W(x) = O(e^{-h_1 x_n})$. Since $v = u$ for sufficiently large x_n, we obtain the assertion of Theorem 11.

References

[1] H. Berestycki, L. Nirenberg, Some qualitative properties of solutions of semilinear equations in cylindrical domains, *Analysis*, ed. by P. Rabinovitz, Academic Press, 1990, 114–164.

[2] V.A.Kondratiev, O.A. Oleinik, On asymptotic behavior of solutions of some nonlinear elliptic equations in unbounded domains, *Partial differential equations and related subjects*, Longman, 1992, 163–195.

[3] S.Agmon, L. Nirenberg, Properties of solutions of ordinary differential equations in Banach space, *Comm .Pure and Appl. Math.* **16**, 1963, 121–239.

[4] A. Pazy, Asymptotic expansions of solutions of ordinary differential equations in Hilbert space, *Arch. Rat. Mech.* **24**, 1967, 193–218.

[5] V.A. Kozlov, V.G. Maz'ya, Estimates and asymptotics of solutions of elliptic boundary value problems in a cone, *Math.Nach.* **137**, 1988, 113–139.

[6] V.A. Kozlov, V.G. Maz'ya, On the asymptotic behaviour of solutions of ordinary differential equations with operator coefficients, Preprint Universitet Lincoping, LITH-MAT-91-49.

[7] D. Gilbarg, N.S. Trudinger, Elliptic partial differential equations of second order, 2nd edition, Springer Verlag, Berlin, 1983.

[8] I. Gohberg, M.G. Krein, Introduction to the theory of nonselfadjoint operators, Moscow, Nauka, 1965.

[9] E. De Giorgi, Sulla differenziabilità e l'analiticà delle estremali degli integrali, *Mem Accad. Sci. Torino*, 1957, 1–19.

[10] V.A. Kondratiev, O.A. Oleinik, Boundary value problems for nonlinear elliptic equations in cylindrical domains, *J.Partial Diff. Eqs.* **6**, N 1, 1993, 10–16.

[11] E.M. Landis, Second order elliptic and parabolic equations, Nauka, Moscow, 1971.

2.3 Asymptotics of solutions of the equation $\Delta u - |u|^{p-1}u = 0$ in cylindrical domains

In this lecture we study in more detail the asymptotics of solutions of the equation $\Delta u - |u|^{p-1}u = 0$, $p > 1$, in a cylindrical domain with the Dirichlet or the Neumann boundary conditions on the lateral part of the boundary of the cylindrical domain. This equation arises in many problems of mathematical physics. We will use here the same notation as in section 1. We set $x = (x_1, \ldots, x_n)$, $x' = (x_1, \ldots, x_{n-1})$. $S(a,b) = \{x : x' \in \omega, a < x_n < b\}$, $\sigma(a,b) = \{x : x' \in \partial\omega, a < x_n < b\}$; ω is a bounded domain in $\mathbf{R}^{n-1} = (x_1, \ldots, x_{n-1})$ with a smooth boundary $\partial\omega$.

Here we study the asymptotic behavior as $x_n \to \infty$ of solutions of the equation

$$\Delta u - |u|^{p-1}u = 0 \quad \text{in } S(0,\infty) \tag{1}$$

with the boundary condition

$$\frac{\partial u}{\partial \nu} = 0 \quad \text{on } \sigma(0,\infty), \tag{2}$$

where ν is the exterior unit normal to $\sigma(0,\infty)$, $p > 1$. In section 1 the problem (1),(2) was also considered and the following theorem was proved.

Theorem 1. *Let $u(x)$ be a solution of equation (1) in $S(0,\infty)$ with the boundary condition (2) on $\sigma(0,\infty)$, $p > 1$. Assume that $u(x)$ attains positive and negative values (oscillates) in any domain $S(k,\infty)$, $k > 0$.*
Then

$$|u(x)| \leqslant Ce^{-\alpha x_n}, \quad \alpha, C = const > 0. \tag{3}$$

First we prove here the following result. The proof is different from the proof of Theorem 4 of section 2.

Theorem 2. *Let $u(x)$ be a positive solution of the problem (1),(2) in $S(0,\infty)$ with $p > 1$. Then*

$$u(x) = M_p(x_n + \gamma)^{\frac{2}{1-p}} + v(x), \tag{4}$$

where

$$|v(x)| \leqslant Ce^{-\beta x_n}, \quad C, \beta, \gamma = const > 0, \tag{5}$$

and

$$M_p = \left(\frac{2(1+p)}{(p-1)^2}\right)^{\frac{1}{p-1}}. \tag{6}$$

We consider classical solutions of the problem (1),(2). This means that $u \in C^2(S(0,\infty)) \cap C^1(\bar{S}(0,\infty))$.

The following theorem is an essential improvement of Theorems 1 and 2 and it is a main result of this lecture.

Theorem 3. *Let $u(x)$ be a solution of equation (1) in $S(0,\infty)$ with boundary condition (2) on $\sigma(0,\infty)$, $p > 1$. Then*

$$u(x) = \pm K(x_n + h)^{\frac{2}{p-1}} + \sum_{i=1}^{m} A_i v_i(x) e^{-\sqrt{\lambda_2} x_n} + o\left(e^{-\sqrt{\lambda_2} x_n}\right), \quad x_n \to \infty,$$
(7)

where the constant K is equal to zero or $K = M_p$, $h = const$, $A_i = const$, $i = 1, \ldots, m$; λ_2 is the second eigenvalue of the eigenvalue problem

$$\Delta v + \lambda v = 0 \quad in \ \omega,$$
(8)

$$\frac{\partial v}{\partial \nu} = 0 \quad on \ \partial \omega,$$
(9)

$v_i(x')$, $i = 1, \ldots, m$, *are the system of orthogonal eigenfunctions, corresponding to λ_2; m is the multiplicity of λ_2. The constants h, A_i ($i = 1, \ldots, m$) depend on $u(x)$.*

Before proving Theorem 3 we prove Theorem 2 and some auxiliary propositions.

Proof of Theorem 2. In section 1 the following inequality is proved:

$$M_p(x_n + \gamma)^{\frac{2}{1-p}} \leqslant u(x) \leqslant M_p x_n^{\frac{2}{1-p}}.$$
(10)

From the inequality (10) it follows that the set of nonnegative numbers δ such that

$$u(x) \leqslant M_p(x_n + \delta)^{\frac{2}{1-p}}$$

for $x_n \geqslant N_\delta$ is not empty. We set $\delta_0 = \sup \delta$.

Consider first the case when $\delta_0 < \infty$. From the definition of δ_0 it follows that $u(x) \leqslant M_p(x_n + \delta_0 - \epsilon)^{\frac{2}{1-p}}$ for any $\epsilon = const > 0$ and $x_n > X_\epsilon$ and in addition

$$u(x^\epsilon) \geqslant M_p(x_n^\epsilon + \delta_0 + \epsilon)^{\frac{2}{1-p}}, \quad x_n^\epsilon \to \infty.$$
(11)

Consider the function $v(x) = u(x) - M_p(x_n + \delta_0)^{\frac{2}{1-p}}$. It is evident that

$$v(x) \leqslant M_p(x_n + \delta_0 - \epsilon)^{\frac{2}{1-p}} - M_p(x_n + \delta_0)^{\frac{2}{1-p}}$$
(12)

for $x_n > X_\epsilon$. It follows from (12) that

$$v(x) \leqslant M_p(x_n + \delta_0)^{\frac{2}{1-p}}\left[\left(1 - \frac{\epsilon}{(x_n + \delta_0)}\right)^{\frac{2}{1-p}} - 1\right].$$

Therefore

$$v_+(x) = o\left(x_n^{\frac{1+p}{1-p}}\right) \quad \text{as } x_n \to \infty, \tag{13}$$

where $v_+(x) = v(x)$ if $v(x) \geqslant 0$, and $v_+(x) = 0$ if $v(x) < 0$. It is easy to see that

$$v_-(x) = \mathcal{O}\left(x_n^{\frac{2}{1-p}}\right) \quad \text{as } x_n \to \infty, \tag{14}$$

where $v_-(x) = v(x)$ if $v(x) \leqslant 0$, and $v_-(x) = 0$, if $v(x) > 0$.

We define the function $Z(x)$ in the following way:

$$Z(x)(x_n + \delta_0)^{\frac{2}{1-p}} = u(x) - M_p(x_n + \delta_0)^{\frac{2}{1-p}}. \tag{15}$$

The function Z satisfies the equation

$$\Delta Z + 4[(1-p)(x_n + \delta_0)]^{-1}\frac{\partial Z}{\partial x_n} + \frac{2(p+1)}{(1-p)^2(x_n + \delta_0)^2}Z$$
$$+ A_p(x, Z)[Z(x_n + \delta_0)^{\frac{2}{1-p}}]^{-1}Z = 0, \tag{16}$$

where

$$A_p(x, Z)$$
$$= 2M_p(1-p)^{-2}(1+p)(x_n + \delta_0)^{-2+\frac{2}{1-p}} - [M_p(x_n + \delta_0)^{\frac{2}{1-p}}$$
$$+ Z(x_n + \delta_0)^{\frac{2}{1-p}}]^p.$$

For the function Z we have an estimate from below,

$$Z(x) = u(x)(x_n + \delta_0)^{-\frac{2}{1-p}} - M_p$$
$$\geqslant M_p\left[\left(\frac{x_n + \gamma}{x_n + \delta_0}\right)^{\frac{2}{1-p}} - 1\right]$$
$$= \mathcal{O}(x_n^{-1}),$$

and an estimate from above,

$$Z(x) \leqslant M_p\left[\left(1 - \frac{\epsilon}{x_n + \delta_0}\right)^{\frac{2}{1-p}} - 1\right] = o(x_n^{-1}). \tag{17}$$

Therefore we get

$$|Z(x)| \leqslant \mathcal{O}(x_n^{-1}), \quad Z_+(x) \leqslant o(x_n^{-1}), \quad x_n \to \infty. \tag{18}$$

Let us prove that $Z_-(x^\epsilon) = o\left(\frac{1}{x_n^\epsilon}\right)$, where x^ϵ is defined by (11). According to (11) we have

$$Z(x^\epsilon) \geqslant M_p \left[1 - \frac{\epsilon}{x_n^\epsilon + \delta_0}\right]^{\frac{2}{1-p}} - M_p \geqslant \frac{2\epsilon M_p}{(1-p)(x_n^\epsilon + \delta_0)}. \tag{19}$$

It follows from (19) that $Z_-(x^\epsilon) = o\left(\frac{1}{x_n^\epsilon}\right)$. Let us estimate the coefficients of the equation (16) for Z. We have

$$A_p(x, Z) = M_p \frac{2(1+p)}{(1-p)^2}(x_n + \delta_0)^{\frac{2p}{1-p}} \left(1 - \left(1 - \frac{Z}{M_p}\right)^p\right).$$

Thus the function $Z(x)$ satisfies the equation

$$\Delta Z + a_1(x_n)\frac{\partial Z}{\partial x_n} + a_2(x_n)Z = 0,$$

where

$$a_1(x_n) \to 0, \ a_2(x_n) \to 0, \ a_1(x_n) = \mathcal{O}(x_n^{-1}), \ a_2 = \mathcal{O}(x_n^{-2}) \ \text{as } x_n \to \infty. \tag{20}$$

We shall prove that $|Z(x)| = o(x_n^{-1})$ as $x_n \to \infty$. We have $Z_+(x) = o(x_n^{-1})$ according to (18). Consider the function

$$w(x) = -Z(x) + \frac{\epsilon}{x_n}.$$

It is easy to see that

$$w(x) \geqslant \frac{\epsilon}{2x_n} \quad \text{for } x_n \geqslant X_\epsilon, \tag{21}$$

since $Z_+(x) = o(x_n^{-1})$. For the sequence x_n^ϵ we have

$$w(x_n^\epsilon) \leqslant 2\epsilon(x_n^\epsilon)^{-1}. \tag{22}$$

The function w satisfies the equation

$$\Delta w + a_1 w_{x_n} + a_2 w - 2\epsilon x_n^{-3} + a_1 \epsilon x_n^{-2} - a_2 \epsilon x_n^{-1} = 0 \tag{23}$$

in $S(0, \infty)$ with the boundary condition $\dfrac{\partial w}{\partial \nu} = 0$ on $\sigma(0, \infty)$. Let w_0 be a solution of the equation (23) in $S(T-2, T+2)$ with the boundary conditions

$$w_0(x', T-2) = 0, \ w_0(x', T+2) = 0, \ \frac{\partial w_0}{\partial \nu} = 0 \text{ on } \sigma(T-2, T+2). \quad (24)$$

The solution w_0 exists for T sufficiently large, since $a_2(x_n) \to 0$ as $x_n \to \infty$. Define

$$W = w - w_0$$

in $S(T-2, T+2)$. We have

$$\Delta W + a_1 W_{x_n} + a_2 W = 0 \text{ in } S(T-2, T+2) , \quad (25)$$
$$\frac{\partial W}{\partial \nu} = 0 \text{ on } \sigma(T-2, T+2) ,$$
$$W = w \text{ for } x_n = T+2 \text{ and } x_n = T-2 . \quad (26)$$

In order to prove that $W > 0$ in $S(T-2, T+2)$ we use the following well-known proposition:

Proposition 1. *Let $u(x)$ be a solution of the equation*

$$\Delta u + \sum_{j=1}^{n} a^j u_{x_j} + au = f$$

in the domain $S(T-2, T+2)$ with the boundary condition $\dfrac{\partial u}{\partial \nu} = 0$ on $\sigma(T-2, T+2)$ and with $a(x)$ sufficiently small in $S(T-2, T+2)$. Then

$$|u(x)| \leqslant C \left[\sup_{S(T-2,T+2)} |f| + \max_{\substack{x_n = T-2 \\ x_n = T+2}} |u| \right] , \quad (27)$$

C being a constant which does not depend on u.

From (21) and the estimate (27), applied to the solution w_0 of the problem (23), (24), we get

$$W(x) \geqslant \frac{\epsilon}{2x_n} - \mathcal{O}(x_n^{-3}) \geqslant \frac{\epsilon}{4x_n} \text{ in } S(T-2, T+2) .$$

For the function W which is a solution of the problem (25),(26) we apply the Harnack inequality: for any positive solution W of the problem (25),(26) in $S(T-2, T+2)$ the inequality

$$W(x) \leqslant C W(x^0) \quad (28)$$

is valid, where $x \in S(T-1, T+1)$, $x^0 \in S(T-1, T+1)$. The constant C does not depend W, x^0, x.

This proposition can be proved in the same way as the Harnack theorem is proved in [1].

It follows from (28) that

$$W(x) \leqslant CW(x^\epsilon) \leqslant C_1 \epsilon x_n^{-1} , \tag{29}$$

since $W > 0$ in $S(T-2, T+2)$ and $W(x^\epsilon) = w(x^\epsilon) - w_0(x^\epsilon) \leqslant 2\epsilon(x_n^\epsilon)^{-1} + \mathcal{O}((x_n^\epsilon)^{-3}) \leqslant 3\epsilon(x_n^\epsilon)^{-1}$.

Therefore we get $w = W + w_0$ and

$$|w(x)| \leqslant C_2 \epsilon x_n^{-1} + C_3 x_n^{-3} \leqslant o(x_n^{-1}) \text{ for } x_n > X_\epsilon , \tag{30}$$

by (29) and the estimate (27) for w_0. Since $Z = \epsilon x_n^{-1} - w(x)$, we obtain from (30) that

$$Z(x) \geqslant \epsilon x_n^{-1} - o(x_n^{-1}) = o(x_n^{-1}) , Z_-(x) = o(x_n^{-1}) . \tag{31}$$

From (31) and (18) we get

$$|Z(x)| = o(x_n^{-1}) .$$

From (31) we have

$$\begin{aligned} v(x) &= u(x) - M_p(x_n + \delta_0)^{\frac{2}{1-p}} \\ &= Z(x)(x_n + \delta_0)^{\frac{2}{1-p}} = o\left(x_n^{\frac{1+p}{1-p}}\right) . \end{aligned} \tag{32}$$

Set

$$\theta(x_n) = M_p(x_n + \delta_0)^{\frac{2}{1-p}}.$$

For $v(x)$ we have the equation

$$\Delta v = g(x)v,$$

where

$$g(x) = (|u|^{p-1}u - \theta^p)(u - \theta)^{-1}. \tag{33}$$

It is evident that $g(x) = p|\tilde{u}|^{p-1}$, where

$$\tilde{u} = \theta + o(x_n^\lambda), \quad \lambda = (p+1)(1-p)^{-1}, \quad \text{as } x_n \to \infty,$$

according to (32). Therefore we have

$$
\begin{aligned}
g(x) &= p(M_p)^{p-1} x_n^{-2} (1 + \delta_0 x_n^{-1})^{-2} (1 + o(x_n^{-1})) \\
&= p(M_p)^{p-1} x_n^{-2} + \mathcal{O}(x_n^{-3}).
\end{aligned}
$$

Let us consider the function $V(x) = S(x_n) v(x)$, where $S(t) = 0$ for $t < \tau$, $\tau = const > 0$, $S(t) = 1$ for $t > \tau + 1$, and $S \in C^\infty(\mathbf{R}^1)$. The function V satisfies the equation

$$
\Delta V - g_1 V = f \text{ in } S(-\infty, +\infty), \tag{34}
$$

$\dfrac{\partial V}{\partial \nu} = 0$ on $\sigma(-\infty, +\infty)$. In (34) f has a compact support and $g_1 = g$ for $x_n > \tau + 1$. In section 1 (see also section 2) it is proved that equation (34) has in $S(-\infty, +\infty)$ a solution V_0 with the following properties:

$$
V_0(x) = \mathcal{O}(e^{-\alpha x_n}) \text{ as } x_n \to \infty, \tag{35}
$$
$$
V_0(x) = a x_n + b + \mathcal{O}(e^{\alpha x_n}) \text{ as } x_n \to -\infty. \tag{36}
$$

Here a, b, α are constants and $\alpha > 0$.

Further, for $V_0(x)$ the following estimate is valid:

$$
\int\limits_{S(-\infty,+\infty)} \sum_{|\beta| \leqslant 2} |\mathcal{D}^\beta V_0|^2 e^{2\alpha x_n} \, dx
$$
$$
\leqslant C_1 \int\limits_{S(-\infty,+\infty)} |f|^2 e^{2\alpha x_n} \, dx
$$
$$
\leqslant C_2 , \qquad C_1, C_2 = const. \tag{37}
$$

It is well-known that for the solution of equation (34) with the boundary condition $\dfrac{\partial V_0}{\partial \nu} = 0$ on $\sigma(-\infty, +\infty)$ for $t > \tau + 1$ we have

$$
\max_{x \in S(t-1,t+1)} |V_0(x)|^2 \leqslant C_3 \int\limits_{S(t-2,t+2)} |V_0|^2 dx ,
$$

where C_3 is a constant not depending on t, V_0. (This is the De Giorgi type estimate (see [2]–[4]).

For the solution $V_0(x)$ we have

$$
\max_{x \in S(t-1,t+1)} |V_0(x)|^2 \leqslant C \int\limits_{S(t-2,t+2)} |V_0(x)|^2 dx
$$

$$\leqslant Ce^{-2\alpha(t-2)} \int\limits_{S(t-2,t+2)} |V_0(x)|^2 e^{2\alpha x_n} dx$$

$$\leqslant C_1 e^{-2\alpha t}.$$

Therefore

$$|V_0(x)|^2 \leqslant C_2 e^{-2\alpha x_n}, \qquad C, C_1, C_2 = const > 0. \tag{38}$$

Let us consider $V_1 = V - V_0$. It follows from (32) and (38) that $V_1 = o(x_n^\lambda)$ as $x_n \to \infty$ and according to (36) $V_1(x) = ax_n + b + \mathcal{O}\left(e^{\alpha x_n}\right)$ as $x_n \to -\infty$, a, b, $\alpha = const$, $\alpha > 0$. We shall prove that $a = 0$, $b = 0$. Suppose that $a < 0$. For V_1 we have the equation

$$\Delta V_1 - g_1 V_1 = 0 \text{ in } S(-\infty, +\infty).$$

Let us consider the equation

$$Y''(x_n) - h(x_n)Y(x_n) = 0, \quad -\infty < x_n < +\infty, \tag{39}$$

where

$$h(x_n) = p(M_p)^{p-1} x_n^{-2}.$$

It is easy to verify that $Y(x_n) = x_n^\lambda$, where $\lambda = (p+1)(1-p)^{-1}$, is a solution of equation (39) for $x_n > 0$. Since $a < 0$ and $V_1(x) \to 0$ as $x_n \to \infty$, we have $V_1 > 0$ in $S(-\infty, +\infty)$. Next, introduce the function

$$Y_1(x_n) = Y(x_n) + kx_n^{\lambda - \frac{1}{2}}, \quad k = const > 0.$$

It is easy to see that

$$\Delta Y_1 - g_1 Y_1 \geqslant 0 \text{ and } Y_1(x_n) > 0 \text{ in } S(0, \infty),$$

since $\Delta Y_1 - g_1 Y_1 = kC_0 x_n^{\lambda - \frac{5}{2}} + \mathcal{O}\left(x_n^{-3+\lambda}\right) \geqslant 0$, if $x_n > 0$, k is sufficiently large and $C_0 = const > 0$. Let us consider the function

$$E = mV_1 - Y_1.$$

We have $\Delta E - g_1 E \leqslant 0$ in $S(1, \infty)$, $E(x) \to 0$ as $x_n \to \infty$ and $E(x', 1) > 0$, if m is sufficiently large, since $V_1 > 0$ in $S(-\infty, +\infty)$. The function $E(x)$ cannot have negative values in $S(1, \infty)$, since $E(x)$ cannot attain a negative minimum in $S(1, \infty)$, on $\sigma(1, \infty)$, and for $x_n = 1$. Therefore $E(x_n) \geqslant 0$ in $S(1, \infty)$ and

$$mV_1 \geqslant x_n^\lambda + kx_n^{\lambda - \frac{1}{2}} \text{ in } S(1, \infty). \tag{40}$$

The inequality (40) contradicts $V_1(x) = o(x_n^\lambda)$ as $x_n \to \infty$. In the same way we can get a contradiction if we suppose that $a > 0$ or $b \neq 0$. Thus $V_1(x) \to 0$ as $x_n \to -\infty$. According to the maximum principle $V_1 \equiv 0$ in $S(-\infty, +\infty)$. We have $V = v$ for $x_n > \tau + 1$, $V_0(x) = \mathcal{O}(e^{-\alpha x_n})$ as $x_n \to \infty$ and $u(x) = v(x) + M_p(x_n + \delta_0)^{\frac{2}{1-p}}$. The theorem is proved for the case $\delta_0 < \infty$.

Consider the case $\delta_0 = \infty$. Since $u > 0$, according to (10) $u(x) \geqslant M_p(x_n + \gamma)^{\frac{2}{1-p}}$, $\gamma = const > 0$. Therefore, for any $\delta > 0$ which satisfies the inequality $u(x) \leqslant M_p(x_n + \delta)^{\frac{2}{1-p}}$, we have $\delta < \gamma$. This implies that $\delta_0 < \infty$. Theorem 2 is proved.

We use Theorem 2 for the proof of the main result, Theorem 3. For this we also need the following lemmas.

Lemma 1. *Assume that $u(t)$ is a solution of the equation*

$$\frac{d^2u}{dt^2} - \frac{k^2}{t^2}u = f(t), \quad 0 < t_1 < t < \infty, \quad k = const \geqslant 0, \qquad (41)$$

$|f(t)| \leqslant C_1 e^{-\alpha_1 t}$, $|u(t)| \leqslant C_2 e^{-\alpha_2 t}$. *Then*

$$|u(t)| \leqslant C_3 e^{-\alpha_1 t}, \quad C_1, C_2, C_3 = const > 0,$$

where C_1, C_2, C_3 and α_1, α_2 are positive constants.

Proof. It is easy to verify that the function

$$u_1(t) = t_3^r \int\limits_t^\infty s^{r_2} \left(\int\limits_s^\infty \tau^{r_1} f(\tau) d\tau \right) ds$$

is a solution of the equation (41), if $r_3 + r_2 + r_1 = 0$, $r_3(r_3 - 1) - k^2 = 0$, $2r_3 + r_2 = 0$. If $k = 0$, we set $r_1 = r_2 = r_3 = 0$. It can be easily proved that $|u_1(t)| \leqslant C e^{-\alpha_1 t}$, since $\tau_1 + \tau_2 + \tau_3 = 0$. The function $y(t) = u(t) - u_1(t)$ is a solution of the equation

$$y'' - \frac{k}{t^2 y} = 0$$

and

$$|y(t)| \leqslant C_4 e^{-\alpha t}, \quad \alpha = \min\{\alpha_1, \alpha_2\}. \qquad (42)$$

This equation has two solutions $y_1(t) = t^{q_1}$, $y_2(t) = t^{q_2}$, where q_1, q_2 are roots of the equation $q^2 - q = k^2$. Therefore, the only solution which satisfies the inequality $|y(t)| \leqslant C_4 e^{-\alpha t}$ is $y(t) \equiv 0$. We have $u(t) = u_1(t)$ and hence $|u(t)| \leqslant Ce^{-\alpha t}$. The lemma is proved.

Lemma 2. *Let $u(x)$ be a solution of the equation*

$$\Delta u - \frac{k^2}{x_n^2} u = f(x) \text{ in } S(1, \infty), \tag{43}$$

$$\frac{\partial u}{\partial \nu} = 0 \text{ on } \sigma(1, \infty), \tag{44}$$

where

$$\left. \begin{array}{l} k \geqslant 0, \ |f(x)| \leqslant C_1 e^{-\gamma_1 x_n}, \\ |u(x)| \leqslant C_2 e^{-\gamma_2 x_n}, \\ 0 < \gamma_2 < \gamma_1 < \sqrt{\lambda_2}. \end{array} \right\} \tag{45}$$

Then

$$|u(x)| \leqslant C_3 e^{-\gamma_1 x_n}, \ C_j = const. \tag{46}$$

Proof. We set

$$u(x) = v(x) + v_1(x_n), \tag{47}$$

where

$$v_1(x_n) = \frac{1}{\text{meas } \omega} \int_\omega u(x) dx', \quad \int_\omega v(x) dx' = 0,$$

and

$$f(x) = F(x) + F_1(x_n), \tag{48}$$

$$F_1(x) = \frac{1}{\text{meas } \omega} \int_\omega f(x) dx', \quad \int_\omega F(x) dx' = 0.$$

Integrating equation (43) over ω, we get

$$v_1'' - \frac{k^2}{x_n^2} v_1 = F_1(x_n), \tag{49}$$

$$\Delta v - \frac{k^2}{x_n^2} v = F(x). \tag{50}$$

From the well-known inequality

$$|\nabla u(x)|^2 \leqslant C_4 \left[\max_{S(T-2,T+2)} |f| + \max_{S(T-2,T+2)} |u| \right], \ x \in S(T-1, T+1), \tag{51}$$

where the constant C_4 does not depend on u and $T > 2$, it follows that

$$|\nabla u(x)| \leqslant C_5 e^{-\gamma_2 x_n}, \quad |\nabla v(x)| + |v(x)| \leqslant C_6 e^{-\gamma_2 x_n}. \qquad (52)$$

Multiplying equation (50) by $v(x)$ and integrating it over $S(t, \infty)$, after integration by parts we get

$$\begin{aligned}
T(t) &\equiv \int\limits_{S(t,\infty)} |\nabla v|^2 \mathrm{d}x \\
&= -\int\limits_{\omega(t)} v \frac{\partial v}{\partial x_n} \mathrm{d}x' - \int\limits_{S(t,\infty)} v F \mathrm{d}x - \int\limits_{S(t,\infty)} \frac{k^2}{x_n^2} v^2 \mathrm{d}x, \qquad (53)
\end{aligned}$$

where $\omega(t) = S(0, \infty) \cap \{x : x_n = t\}$. Since $\int\limits_{\omega(t)} v \mathrm{d}x' = 0$, we can apply the Poincaré inequality and also the elementary inequality $\sqrt{a}\sqrt{b} \leqslant \frac{1}{2}(a + b)$, to obtain from (53) that

$$\begin{aligned}
T(t) &\leqslant \frac{1}{\sqrt{\lambda_2}} \left(\int\limits_{\omega(t)} \sum_{j=1}^{n-1} \left(\frac{\partial v}{\partial x_j} \right) \mathrm{d}x' \right)^{1/2} \left(\int\limits_{\omega(t)} \left(\frac{\partial v}{\partial x_j} \right) \mathrm{d}x' \right)^{1/2} \\
&\quad + C_\epsilon \int\limits_{S(t,\infty)} |F|^2 \mathrm{d}x + \epsilon \int\limits_{S(t,\infty)} v^2 \mathrm{d}x \\
&\leqslant \frac{1}{2} \lambda_2^{-1/2} \int\limits_{\omega(t)} |\nabla v|^2 \mathrm{d}x' + C_\epsilon \int\limits_{S(t,\infty)} |F|^2 \mathrm{d}x \\
&\quad + \epsilon \int\limits_{S(t,\infty)} |v|^2 \mathrm{d}x. \qquad (54)
\end{aligned}$$

It follows from (54) that

$$T(t) \leqslant -\frac{1}{2\sqrt{\lambda_2}} T'(t) + C_\epsilon \int\limits_{S(t,\infty)} |F|^2 \mathrm{d}x + \epsilon \frac{1}{\lambda_2} \int\limits_{S(t,\infty)} |\nabla v|^2 \mathrm{d}x,$$

$$(2\sqrt{\lambda_2} - 2\epsilon \lambda_2^{-1/2}) T(t) + T'(t) \leqslant 2\sqrt{\lambda_2} C_\epsilon \int\limits_{S(t,\infty)} |F|^2 \mathrm{d}x.$$

Let us take ϵ such that $2\lambda_2^{1/2} - 2\epsilon \lambda_2^{-1/2} = 2\gamma_1 + \delta$, where $\delta = const > 0$. Then we have

$$2\gamma_1 T + \delta T + T' \leqslant C_7 e^{-2\gamma_1 t}, \quad C_7 = const > 0. \qquad (55)$$

If the constant A is sufficiently large, we find from (55) that

$$(Ae^{-2\gamma_1 t} - T)' + (2\gamma_1 + \delta)(Ae^{-2\gamma_1 t} - T) = \delta Ae^{-2\gamma_1 t}$$

$$- (T' + (2\gamma_1 + \delta)T) \geqslant \delta Ae^{-2\gamma_1 t} - C_7 e^{-2\gamma_1 t} > 0, \tag{56}$$

$$(Ae^{-2\gamma_1 t} - T)|_{t=t_1} > 0. \tag{57}$$

From (56) and (57) we obtain

$$T \leqslant Ae^{-2\gamma_1 t} \text{ for } t \geqslant t_1. \tag{58}$$

By the Poincaré inequality and (58) we have

$$\int\limits_{S(t,\infty)} |v|^2 \mathrm{d}x \leqslant A_1 e^{-2\gamma_1 t}, \ A_1 = const. \tag{59}$$

From the E. De Giorgi estimate ([2],[4]) and (59) it follows that

$$|v(x)| \leqslant C_8 e^{-\gamma_1 x_n}. \tag{60}$$

We can apply Lemma 1 to $v_1(x)$ and get

$$|v_1(x_n)| \leqslant C_9 e^{-\gamma_1 x_n}. \tag{61}$$

The inequality (60) and (61) yield (46) and the lemma is proved.

Lemma 3. *Let $u(x)$ be a solution of the equation (43) with the boundary condition (44) and*

$$|f(x)| \leqslant C_1 e^{-\gamma_1 x_n}, \ |u(x)| \leqslant C_2 e^{-\gamma_2 x_n}, \ \gamma_1 > \sqrt{\lambda_2}. \tag{62}$$

Then

$$|u(x)| \leqslant C_3 e^{-\sqrt{\lambda_2} x_n}, \tag{63}$$

$C_j = const, \ j = 1, 2, 3.$

Proof. From (62) and Lemma 2 it follows that

$$|u(x)| \leqslant C_\epsilon e^{(-\sqrt{\lambda_2}+\epsilon)x_n}, \ C_\epsilon = const,$$

for any $\epsilon = const > 0$. As in the proof of Lemma 2 we consider the solution $u(x)$ in the form (47). From (53) we get for $v(x)$

$$\begin{aligned}
T(t) &\leqslant \frac{1}{2\sqrt{\lambda_2}} \int\limits_{\omega(t)} |\nabla v|^2 \mathrm{d}x' + \int\limits_{S(t,\infty)} |F| \, |v| \mathrm{d}x \\
&\leqslant -\frac{1}{2\sqrt{\lambda_2}} T'(t) + C_4 e^{(-\gamma_1 - \sqrt{\lambda_2}+\epsilon)x_n} \\
&\leqslant -\frac{1}{2\sqrt{\lambda_2}} T'(t) + C_5 e^{(-2\sqrt{\lambda_2}-\delta_1)x_n},
\end{aligned} \tag{64}$$

where $\delta_1 = const > 0$. It follows from (64) that

$$\left(e^{2\sqrt{\lambda_2}x_n}T(x_n)\right)' \leqslant C_6 e^{-\delta_1 x_n}, \ e^{2\sqrt{\lambda_2}x_n}T(x_n) \leqslant C_4.$$

Then applying the Poincaré inequality, the E. De Giorgi theorem and Lemma 1 as in the proof of Lemma 2 we obtain (63).

Lemma 4. *Suppose that $u(x)$ is a solution of the problem (1),(2) and the conditions of Theorem 1 are satisfied. Then*

$$u(x) = \mathcal{O}\left(e^{-\sqrt{\lambda_2}x_n}\right), \ x_n \to \infty. \tag{65}$$

If the conditions of Theorem 2 are satisfied, then

$$u(x) = M_p(x_n + \gamma)^{\frac{2}{1-p}} + \mathcal{O}\left(e^{-\sqrt{\lambda_2}x_n}\right), \ x_n \to \infty, \tag{66}$$

where γ is a constant which depends on $u(x)$.

Proof. Let α be the constant in (3). Consider the sequence

$$\sigma_1 = \alpha, \ \sigma_2 = p\alpha, \ldots, \ \sigma_k = p^{k-1}\alpha.$$

If $\alpha > \sqrt{\lambda_2}$, then (65) is proved. Suppose that $\sigma_m \leqslant \sqrt{\lambda_2}, \sigma_{m+1} > \sqrt{\lambda_2}$. From equation (1) we have

$$\Delta u = |u|^{p-1}u = f(x), \ |f(x)| \leqslant C_1 e^{-\sigma_2 x_n}. \tag{67}$$

Successively applying Lemma 2 we obtain that $|u(x)| \leqslant C_m e^{(-\sigma_m+\epsilon)x_n}$ for any $\epsilon = const > 0$. Therefore we get

$$\Delta u = |u|^{p-1}u = f(x), \ |f(x)| \leqslant C_{m+1}e^{(-p\sigma_m+\epsilon p)x_n} = C_{m+1}e^{(-\sigma_{m+1}+\epsilon p)x_n}$$

where $\sigma_{m+1} - p\epsilon > \sqrt{\lambda_2}$, if ϵ is sufficiently small. Then (65) follows from Lemma 3.

Assume that (4) is valid for $u(x)$. We set

$$u(x) = M_p(x_n + \gamma)^{\frac{2}{1-p}} + w(x).$$

From (1) we get

$$\Delta w = [w + M_p(x_n + \gamma)^{\frac{2}{1-p}}]^p - \Delta(M_p(x_n + \gamma)^{\frac{2}{1-p}}). \tag{68}$$

We can assume that $\gamma = 0$ and $x_n > 1$. From (68) it is easy to obtain

$$\Delta w - \frac{2p(1+p)}{(p-1)^2}\frac{w}{x_n^2} = f(x), \tag{69}$$

where $|f(x)| \leqslant Cw^2$, $C = const$. Consider the sequence

$$\sigma_1 = \beta, \ \sigma_2 = 2\beta, \ldots, \ \sigma_m = 2^{m-1}\beta, \ \sigma_m \leqslant \sqrt{\lambda_2}, \ \sigma_{m+1} > \sqrt{\lambda_2},$$

β is the constant in (5). From (5) it follows that $|w(x)| \leqslant C_1 e^{-\sigma_1 x_n}$ and therefore $|f(x)| \leqslant C_2 e^{-\sigma_2 x_n}$. Again applying Lemma 2 successively we obtain $|u(x)| \leqslant C_\epsilon e^{-(\sigma_m - \epsilon)x_n}$ and $|f(x)| \leqslant C e^{-2(\sigma_m - \epsilon)x_n}$, where $2\sigma_m - 2\epsilon > \sqrt{\lambda_2}$. Then the estimate (66) follows from Lemma 3.

Lemma 5. *The equation*

$$y'' - \alpha^2 y = f(t) \tag{70}$$

where $\alpha = const > 0$, $|f(t)| \leqslant C_1 e^{-\beta t}$, $\beta > \alpha$, $t > 0$, *has a unique solution such that*

$$|y(t)| \leqslant C_2 e^{-\beta t}, \quad \sup |y(t)e^{\beta t}| \leqslant C_3 \sup |f(t)e^{\beta t}|. \tag{71}$$

Proof. It is easy to see that the function

$$z(t) = e^{\alpha t} \int_t^\infty e^{-2\alpha s} \left(\int_s^\infty f(\tau)e^{\alpha \tau} d\tau \right) ds$$

is a solution of equation (70), which satisfies the conditions (71). If $Z_1(t)$ and $Z_2(t)$ are two solutions satisfying (71), then $Z_0 = Z_1 - Z_2$ satisfies the equation $Z_0'' - \alpha^2 Z_0 = 0$ and $|Z_0(t)| \leqslant C_4 e^{-\beta t}$. Since $\beta > \alpha$, $Z_0 \equiv 0$ for $t > 0$.

Lemma 6. *The equation*

$$y'' - \alpha^2 y = g(t)y + f(t), \tag{72}$$

where $|f(t)| \leqslant C_1 e^{-\beta t}$, $\alpha > 0$, $\beta > \alpha$, *has a unique solution which satisfies conditions (71), if* $|g(t)| < \epsilon$ *and* ϵ *is sufficiently small.*

Proof. The equation (72) is equivalent to the integral equation

$$y(t) = e^{\alpha t} \int_t^\infty e^{-2\alpha s} \left(\int_s^\infty e^{\alpha \tau} f(\tau) d\tau \right) ds$$
$$+ e^{\alpha t} \int_t^\infty e^{-2\alpha s} \left(\int_s^\infty g(\tau)y(\tau)d\tau \right) ds. \tag{73}$$

The existence of a solution of (73) with the properties (71) can be proved by the method of successive approximations.

The uniqueness of such a solution follows from the fact that any nontrivial solution $Z(t)$ of the equation $y'' - \alpha^2 y - g(t)y = 0$ satisfies the inequality $Z(t) \geqslant C_2 e^{-(\alpha+\delta)t}$, if ϵ is sufficiently small, $\delta = const > 0$. Indeed, let us take C_2 such that $Z(0) \geqslant C_2$ and δ such that $(\alpha + \delta)^2 - \alpha^2 - g < 0$. Then $v(t) = Z(t) - C_2 e^{-(\alpha+\delta)t} \geqslant 0$ for $t \geqslant 0$, since v cannot attain a negative minimum by virtue of the inequality $v'' - (\alpha^2 + g)v < 0$ and $v \to 0$ as $t \to \infty$, $v(0) \geqslant 0$, $|g(t)| \leqslant \epsilon$.

Lemma 7. *Let $y(t)$ be a solution of the equation*

$$y'' - \alpha^2 y - \frac{k^2}{t^2}y = f(t) \tag{74}$$

where $\alpha \neq 0$, $k \geqslant 0$, $|f(t)| \leqslant C_1 e^{-\beta t}$, $\beta > \alpha$, $|y(t)| \leqslant C_2 e^{-\alpha_1 t}$, $\alpha_1 < \beta$. Then

$$y(t) = C_3 e^{-\alpha t} + o(e^{-\alpha t}) , \qquad t \to \infty . \tag{75}$$

Proof. According to Lemma 6 the equation (74) has a solution $y_1(t)$ such that $|y_1(t)| \leqslant C_4 e^{-\beta t}$. The function $Y_1(t) = y(t) - y_1(t)$ satisfies the equation

$$Y'' - \alpha^2 Y - \frac{k^2}{t^2}Y = 0. \tag{76}$$

It is known [5] that any solution of the equation (76) has the form

$$Y(t) = C_4(1 + o(1))e^{\alpha t} + C_5(1 + o(1))e^{-\alpha t} \text{ as } t \to \infty , \tag{77}$$

where C_4, C_5 are constants. Since $Y_1(t) \to 0$ as $t \to \infty$, we have $C_4 = 0$ for $Y_1(t)$. Therefore, for $y(t)$ (75) is valid.

Lemma 8. *Assume that $u(x)$ is a solution of the equation*

$$\Delta u - \frac{k^2}{x_n^2}u = f(x) \text{ in } S(1,\infty) \tag{78}$$

with the boundary condition

$$\frac{\partial u}{\partial \nu} \text{ on } \sigma(1,\infty) , \tag{79}$$

where $|u(x)| \leqslant C_1 e^{-\sqrt{\lambda_2}x_n}$, $|f(x)| \leqslant C_2 e^{-\beta x_n}$, $\beta > \sqrt{\lambda_2}$. Then

$$u(x) = \sum_{i=1}^{m} A_i v_i(x')e^{-\sqrt{\lambda_2}x_n} + o(e^{-\sqrt{\lambda_2}x_n}) , \qquad x_n \to \infty , \tag{80}$$

where $v_1, \ldots, v_m$ is a system of orthonormal eigenfunctions, corresponding to the eigenvalue λ_2 of the problem (8),(9), and m is the multiplicity of λ_2, $A_i = const$, $i = 1, \ldots, m$.

Proof. We set

$$u(x) = \sum_{i=0}^{m} B_i(x_n) v_i(x') + u^* ,$$

$$f(x) = \sum_{i=0}^{m} F_i(x_n) v_i(x') + F^* ,$$

where

$$v_0 = const , \int_{\omega} u^* v_i dx' = 0 ,$$

$$\int_{\omega} F^* v_i dx' = 0 , i = 0, 1, \ldots, m . \tag{81}$$

From equation (78) we have

$$-\lambda_2 \sum_{i=1}^{m} B_i(x_n) v_i + \sum_{i=0}^{m} B_i''(x_n) v_i$$

$$-\frac{k^2}{x_n^2} \sum_{i=0}^{m} B_i(x_n) v_i + \Delta u^* - \frac{k^2}{x_n^2} u^*$$

$$= \sum_{i=0}^{m} F_i(x_n) v_i + F^* . \tag{82}$$

Multiplying (82) by v_i and integrating over ω, after the integration by parts, we obtain the following relations:

$$B_0''(x_n) - \frac{k^2}{x_n^2} B_0(x_n) = F_0(x_n) , \tag{83}$$

$$B_i''(x_n) - \lambda_2 B_i(x_n) - \frac{k^2}{x_n^2} B_i(x_n) = F_i(x_n) , i = 1, \ldots, m , \tag{84}$$

$$\Delta u^* - \frac{k^2}{x_n^2} u^* = F^* \text{ in } S(1, \infty) , \tag{85}$$

$$\frac{\partial u^*}{\partial \nu} = 0 \text{ on } \sigma(1, \infty) . \tag{86}$$

In order to estimate $B_0(x_n)$ we use Lemma 1. Since $|f(x)| \leqslant C_2 e^{-\beta x_n}$, $\beta > \sqrt{\lambda_2}$, and therefore $|F_0(x_n)| \leqslant C_3 e^{-\beta x_n}$, $|u(x)| \leqslant C_1 e^{-\sqrt{\lambda_2} x_n}$, $|B_0(x_n)| \leqslant C_4 e^{-\sqrt{\lambda_2} x_n}$, according to Lemma 1 we have $|B_0(x_n)| \leqslant C_5 e^{-\beta x_n}$. Using Lemma 7 we have

$$B_i(x_n) = A_i e^{-\sqrt{\lambda_2} x_n} + o(e^{-\sqrt{\lambda_2} x_n}), \qquad i = 1, \ldots, m. \tag{87}$$

Let us estimate $u^*(x)$. Multiplying (85) by u^*, integrating it over $S(t, \infty)$, transforming the first integral by integration by parts and taking into account (86), we obtain

$$\begin{aligned} T(t) &\equiv \int_{S(t,\infty)} |\nabla u^*|^2 dx \\ &= -\int_{\omega(t)} u^* \frac{\partial u^*}{\partial x_n} dx' - \int_{S(t,\infty)} \frac{k^2}{x_n^2} (u^*)^2 dx \\ &\quad - \int_{S(t,\infty)} F^* u^* dx \,. \end{aligned} \tag{88}$$

It follows from (88) that

$$T(t) \leqslant \left(\int_{\omega(t)} (u^*)^2 dx' \right)^{\frac{1}{2}} \left(\int_{\omega(t)} \left(\frac{\partial u^*}{\partial x_n} \right) dx' \right)^{\frac{1}{2}} + C_6 \left(\int_{S(t,\infty)} (u^*)^2 dx \right)^{\frac{1}{2}} e^{-\beta t} \,. \tag{89}$$

Since u^* satisfies the condition (81) from the variational theory of eigenvalues we have

$$\int_{\omega(t)} (u^*)^2 dx' \leqslant \lambda_3^{-1} \int_{\omega(t)} \sum_{i=1}^{n-1} \left(\frac{\partial u^*}{\partial x_i} \right)^2 dx' \,, \tag{90}$$

where λ_3 is the third eigenvalue of the problem (8),(9). From (89) and (90) we obtain

$$T(t) \leqslant -\frac{1}{2} \lambda_3^{-\frac{1}{2}} T'(t) + C_7 e^{(-\sqrt{\lambda_2}-\beta)t} \,, \qquad C_7 = const \,.$$

Integrating this inequality from t_0 to t, we get

$$T(t) \leqslant C_8 e^{-2\gamma t} \,,$$

where $\gamma = \min(\sqrt{\lambda_3}, \frac{1}{2}(\sqrt{\lambda_2} + \beta))$. Using the Poincaré inequality and the De Giorgi theorem, as in the proof of Lemma 2, we get

$$|u^*(x)| \leqslant C_9 e^{-\gamma x_n} , \qquad \gamma > \sqrt{\lambda_2} . \tag{91}$$

From the estimate for $B_0(x_n)$ and (87), (91) we obtain (80). The lemma is proved.

Now we proceed to prove Theorem 3.

Proof of Theorem 3. From Lemma 4 it follows that

$$u(x) = \pm K(x_n + h)^{\frac{2}{1-p}} + w ,$$

where $|w(x)| \leqslant C_1 e^{-\sqrt{\lambda_2} x_n}$. From equation (1) we obtain the equation for w of the form

$$\Delta w = \left| w \pm K(x_n + h)^{\frac{2}{1-p}} \right|^{p-1} (w \pm K(x_n + h)^{\frac{2}{p-1}})$$
$$\mp \Delta(K(x_n + h)^{\frac{2}{1-p}}).$$

Using the Taylor formula and the estimate for w we get

$$\Delta w - pK^{p-1}(x_n + h)^{-2}w = \mathcal{F}(x) ,$$

where $|\mathcal{F}(x)| \leqslant C_2 e^{-2\sqrt{\lambda_2} x_n}$ or $|\mathcal{F}(x)| \leqslant C_3 e^{-p\sqrt{\lambda_2} x_n}$, if $K = 0$. Then (7) follows from Lemma 8. This completes the proof.

The approach used in this lecture can be applied to a more general class of nonlinear elliptic equations.

References

[1] E.M. Landis, Second order elliptic and parabolic equations, Moscow, Nauka, 1971.

[2] E. De Giorgi, Sulla differenziabilità e l'analiticità delle estremali degli integrali, *Mem. Accad. Sci. Torino* 1957, 1–19.

[3] J. Moser, A new proof of the De Giorgi theorem concerning the regularity problem for elliptic differential equations, *Comm. Pure and Appl. Math.* **13**, n. 3, 1960, 457–468.

[4] D. Gilbarg, N.S. Trudinger, Elliptic partial differential equations of second order, 2nd edition, Springer Verlag, Berlin, 1983.

[5] R. Bellman, Stability theory of differential equations, McGraw-Hill, New York, 1953.

2.4 Asymptotics of solutions of the equation $\Delta u - e^u = 0$ in cylindrical domains

In this lecture we consider the equation

$$\Delta u - e^u = 0 \tag{1}$$

which is a model equation for a large class of equations which arise in many problems of mathematical physics and geometry. This class of equations forms the subject of a vast literature (see [1], [2] and the references there).

The Dirichlet boundary conditions. We will use the same notation as in section 3. We consider the equation (1) in the cylindrical domain $S(0, \infty)$ with the boundary conditions

$$u = 0 \text{ on } \sigma(0, \infty). \tag{2}$$

Our aim is to study the behavior of solutions of the problem (1),(2) as $x_n \to \infty$.

Theorem 1. *Let $u(x)$ be a solution of the equation (1) with the boundary condition (2). Then either*

$$u(x) = CZ(x')e^{\sqrt{\lambda_1}x_n} + \mathcal{O}(1), \qquad C = const < 0, x_n \to \infty, \tag{3}$$

or

$$u(x) = u_0(x') + \mathcal{O}(e^{-\sqrt{\lambda_1}x_n}), x_n \to \infty, \tag{4}$$

where λ_1 is the smallest eigenvalue of the eigenvalue problem

$$\Delta u + \lambda_1 u = 0 \text{ in } \omega, \tag{5}$$

$$u = 0 \text{ on } \partial\omega, \tag{6}$$

$Z(x') = Z(x_1, \ldots, x_{n-1})$ is an eigenfunction corresponding to the eigenvalue λ_1 and positive in ω, $u_0(x')$ is a solution of the equation

$$\Delta u_0 - e^{u_0} = 0 \text{ in } \omega, \tag{7}$$

$$u_0 = 0 \text{ on } \partial\omega, u_0 \leqslant 0. \tag{8}$$

The proof of Theorem 1 is based on the following lemmas.

Lemma 1. *Let $u(x)$ be a solution of equation (1) with the boundary condition (2). Then*

$$u(x) < 0 \qquad (9)$$

for $x \in S(X, \infty)$, where $X \geqslant 0$ and X does not depend on u.

Proof. Consider the solution $Y(t)$ of the equation

$$Y'' - e^Y = 0, \qquad (10)$$

satisfying the initial condition

$$Y(T) = 0, Y'(T) = 0. \qquad (11)$$

In Lemma 4, which will be proved at the end of this lecture, the existence of such a solution of the problem (10),(11) is proved. This solution $Y(t)$ is defined on the interval $(T - h, T + h)$, $Y(t)$ is a positive function for $T - h < t < T + h$,

$$\lim_{t \to T \pm h} Y(t) = +\infty,$$

where h is a finite positive constant. Let $u(x) - Y(x_n) = v(x)$. The function $v(x)$ satisfies the equation

$$\Delta v - Q(x)v = 0 \text{ in } S(T - h, T + h), \qquad (12)$$

where $Q(x) \geqslant 0$, and the boundary condition $v \leqslant 0$ on $\sigma(T - h, T + h)$, $v \to -\infty$ as $x_n \to T - h$, $x_n \to T + h$. Applying the maximum principle to equation (12) we have

$$v(x) \leqslant 0 \text{ in } S(T - h, T + h).$$

Therefore

$$u(x) \leqslant Y(x_n) \text{ in } S(T - h, T + h),$$
$$u(x', T) \leqslant Y(T) = 0$$

for any $T \geqslant h$, and (9) is valid in $S(X, \infty)$ for $X = h$.

Lemma 2. *Let $u(x) \geqslant 0$ in $S(0, \infty)$,*

$$\Delta u \leqslant 0 \text{ in } S(0, \infty) \text{ and } u = 0 \text{ on } \sigma(0, \infty). \qquad (13)$$

Then

$$\int_{S(0,T)} u dx \leqslant C_1 e^{\alpha T}, \qquad C_1, \alpha, T = const > 0,$$

and α does not depend on u.

Proof. Let us consider the auxiliary function $\psi_1(x') \in C^2(\overline{\omega})$, such that $\psi_1(x') > 0$ for $x' \in \omega$ and $\psi_1(x') = \rho^2(x', \partial\omega)$ in a neighborhood of $\partial\omega$, where $\rho(x, \partial\omega)$ is the distance from x' to $\partial\omega$. It is easy to see that

$$\psi_1 = 0 \text{ on } \partial\omega, \text{grad } \psi_1 = 0 \text{ on } \partial\omega,$$

$\Delta\psi_1 = 2\rho\Delta\rho + 2|\nabla\rho|^2 > 0$ on $\partial\omega$, since $\nabla\rho \neq 0$ on $\partial\Omega$. Therefore $\Delta\psi_1 > 0$ for $x' \in \omega_0$, which is a neighborhood of $\partial\omega$. Setting

$$\psi(x) = x_n^2 \psi_1(x'),$$

it is evident that

$$\Delta\psi = 2\psi_1(x') + x_n^2\Delta\psi_1$$

and $\Delta\psi > 0$ for $x' \in \omega_0$ and $x_n > 0$, $\Delta\psi > 0$, if $x' \in \omega\backslash\omega_0$, $0 < x_n < \epsilon$, since $\psi_1 > C_2 = const > 0$ in $\omega\backslash\omega_0$. We thus have

$$\Delta\psi \geqslant 0 \text{ in } S(0, \epsilon_1) \text{ and}$$
$$\Delta\psi > \delta = const > 0 \text{ in } S(\epsilon/2, \epsilon). \tag{14}$$

Consider the function

$$\phi(x) = \psi(x)\theta(x_n),$$

where $\theta(x_n) \equiv 1$ for $x_n < \frac{3\epsilon}{4}$, $\theta(x_n) = 0$ for $x_n \geqslant \epsilon$, and $\theta(x_n) \geqslant 0$, $\theta \in C^\infty(\mathbf{R}^1)$. Multiplying equation(1) by $\theta(x)$, transforming the first integral by integration by parts, we get

$$\int_{S(0,\epsilon)} u\Delta\phi dx \leqslant 0.$$

Therefore

$$\int_{S(0,\epsilon/2)} u\Delta\phi dx + \int_{S(\epsilon/2,3\epsilon/4)} u\Delta\psi dx \leqslant C_3 \int_{S(3\epsilon/4,\epsilon)} u dx.$$

From (14) it follows that

$$\int_{S(\epsilon/2,3\epsilon/4)} u dx \leqslant C_4 \int_{S(3\epsilon/4,\epsilon)} u dx. \tag{15}$$

After the transformation

$$\hat{x}_n = \epsilon - x_n$$

we obtain from (15) that

$$\int_{S(\epsilon/4,\epsilon/2)} u\,dx \leqslant C_5 \int_{S(0,\epsilon/4)} u\,dx.$$

In a similar way we obtain

$$\int_{S(a,a+\epsilon/4)} u\,dx \leqslant C_6 \int_{S(a-\epsilon/4,a)} u\,dx. \tag{16}$$

Therefore, applying inequality (16) K times, we obtain

$$\int_{S\left(\frac{K\epsilon}{4},\frac{(K+1)\epsilon}{4}\right)} u\,dx \;\leqslant\; C_6 \int_{S\left(\frac{\epsilon(K-1)}{4},\frac{K\epsilon}{4}\right)} u\,dx$$
$$\leqslant \; \cdots \leqslant C_6^K \int_{S(0,\frac{\epsilon}{4})} u\,dx \tag{17}$$

for any $K > 0$. From (17) it follows that

$$\int_{S(0,\frac{\epsilon(K+1)}{4})} u\,dx \leqslant KC_6^K \int_{S(0,1/4)} u\,dx,$$

and

$$\int_{S(0,T)} u\,dx \leqslant e^{\alpha T} \int_{S(0,1/4)} u\,dx.$$

Lemma 2 is proved.

Lemma 3. *Let*

$$\Delta u = f(x) \; in \; S(0,\infty), \tag{18}$$

and let ω be a domain with a smooth boundary in $\mathbf{R}^{n-1}$,

$$u = 0 \; on \; \sigma(0,\infty),$$
$$u \geqslant 0 \; in \; S(0,\infty), -1 \leqslant f(x) \leqslant 0.$$

Then

$$u(x) = C_1 e^{\sqrt{\lambda_1}x_n}\varphi_1(x') + \mathcal{O}(1), \tag{19}$$

where $C_1 = const > 0$, $\lambda_1 > 0$ is the smallest eigenvalue of the problem

$$\Delta\varphi + \lambda_1\varphi = 0, x' \in \omega, \varphi = 0 \; on \; \partial\omega,$$

$\varphi_1(x') > 0$ in ω and φ_1 is an eigenfunction, corresponding to the eigenvalue $\lambda = \lambda_1$.

Proof. Consider the function $u_m(x)$ such that

$$\Delta u_m = f(x) \text{ in } S(0, m)$$

$u_m(x) = u(x', 0)$ for $x_n = 0$, $u_m(x) = 0$ for $x \in \sigma(0, m)$ and $u_m(x) = 0$ for $x_n = m$. Also, assume that

$$\int\limits_{S(0,m)} |\nabla u_m|^2 \mathrm{d}x < \infty.$$

We shall prove that the sequence $u_m(x)$ is uniformly bounded with respect to m. Consider the function u^* such that

$$\Delta u^* = -2 \text{ in } \omega, u^* = 0 \text{ on } \partial\omega.$$

It is evident that $u^*(x') > 0$ in ω. The function

$$V = u^*(x') + K - u_m(x), \qquad K = const,$$

satisfies the equation $\Delta V = -2 - f < 0$ and $V > 0$ on $\partial S(0, m)$, if $u(x', 0) < K$. By the maximum principle, $V > 0$ in $S(0, m)$ and therefore

$$u_m(x) \leqslant K + u^*(x') \leqslant K_1, \quad K_1 = const. \tag{20}$$

It is clear that $u_m \geqslant 0$ and (20) implies $|u_m| \leqslant K_1$. From (20) and the properties of the Laplace operator it follows that u_m contains a sub-sequence $u_{m'}$ such that $u_{m'} \to \tilde{u}$ as $m' \to \infty$ uniformly in any $S(0, M)$, $M > 0$. For $\tilde{u}$ we have

$$\Delta \tilde{u} = f \text{ in } S(0, \infty),$$
$$\tilde{u} = 0 \text{ on } \sigma(0, \infty), 0 \leqslant \tilde{u} \leqslant K_1. \tag{21}$$

Setting $v(x) = u(x) - \tilde{u}(x)$, we have

$$\Delta v = 0 \text{ in } S(0, \infty),$$
$$v = 0 \text{ on } \sigma(0, \infty),$$
$$v = 0 \text{ for } x_n = 0.$$

According to (21) and Lemma 2, applied to the function u, we get

$$\int\limits_{S(0,x_n)} v \mathrm{d}x \leqslant C_2 e^{\alpha x_n}, \alpha = const > 0. \tag{22}$$

We set

$$v_k(x_n) = \int\limits_{\omega} v(x', x_n)\varphi_k(x')\mathrm{d}x', \qquad (23)$$

where $\varphi_k(x')$ is an eigenfunction, corresponding to the eigenvalue $\lambda = \lambda_k$, and $0 < \lambda_1 < \lambda_2 \leqslant \cdots \leqslant \lambda_k \leqslant \cdots$, every eigenvalue in this sequence being counted as many times as its multiplicity. It is easy to see that

$$
\begin{aligned}
v_k''(x_n) &= \int\limits_{\omega} \frac{\partial^2 v}{\partial x_n^2}\varphi_k(x')\mathrm{d}x' = -\int\limits_{\omega} \Delta_{x'}v\varphi_k(x')\mathrm{d}x' \\
&= \int\limits_{\omega} v\Delta\varphi_k\mathrm{d}x' = \lambda_k v_k, \qquad 0 < x_n < \infty, \qquad (24) \\
v_k(0) &= 0. \qquad\qquad\qquad\qquad\qquad\qquad\qquad (25)
\end{aligned}
$$

Therefore,

$$v_k(x_n) = C^{(k)}\sin h(\sqrt{\lambda_k}x_n). \qquad (26)$$

From (23) we have

$$
\begin{aligned}
\int\limits_0^{x_n} v_k(x_n)\mathrm{d}x_n &= \int\limits_{S(0,x_n)} v(x)\varphi_k(x')\mathrm{d}x'\mathrm{d}x_n \\
&\leqslant C(k)\int\limits_{S(0,x_n)} v(x)\mathrm{d}x \\
&\leqslant C^1(k)e^{\alpha x_n}. \qquad (27)
\end{aligned}
$$

From (26) and (27) it follows that $v_k = 0$ for $\lambda_k > \alpha^2$. Hence

$$
\begin{aligned}
v(x) &= \sum_{k=1}^{N} v_k(x_n)\varphi_k(x') \\
&= \sum_{k=1}^{N} C^{(k)}\sin h(\sqrt{\lambda_k}x_n)\varphi_k(x'), \qquad (28)
\end{aligned}
$$

where $N < \infty$.

Since $\tilde{u}$ is a bounded function in $S(0,\infty)$ and $u \geqslant 0$ in $S(0,\infty)$, we have

$$|v_-(x)| \leqslant C_3, \qquad C_3 = const, \qquad (29)$$

where $v_-(x) = v(x)$ if $v(x) \leqslant 0$, and $v_-(x) = 0$ if $v(x) > 0$. The equality (28) can be written in the form

$$v(x) = \sum_{k'=1}^{N'} C^{k'} \sin h(\sqrt{\lambda_{k'}}x_n)\widetilde{\varphi}_{k'}(x'), \qquad (30)$$

where $\{\lambda_{k'}\}$ is a sequence such that every eigenvalue is counted only once independently of its multiplicity. It is easy to see from (30) that $N' = 1$, since if $N' > 1$, then $\varphi_{k'}$ with $k' > 1$ changes sign in ω, but this contradicts (29). Therefore, we have

$$u(x) = v(x) + \tilde{u}(x)$$

and, since $\tilde{u}(x)$ is a bounded function,

$$u(x) = C^1 \sin h(\sqrt{\lambda_1}x_n)\varphi_1(x') + \mathcal{O}(1). \qquad (31)$$

The equality (31) proves Lemma 3.

Proof of Theorem 1. We note that the solution $u_0(x)$ of the problem (7),(8) exists, $u_0(x)$ is unique and negative in ω. The uniqueness of $u_0(x)$ follows from the maximum principle. Indeed, let $u_1(x)$ and $u_2(x)$ be solutions of the problem (7),(8). Then for $v = u_1 - u_2$ we have

$$\Delta v - Q(x)v = 0 \text{ in } \omega,$$
$$v = 0 \text{ on } \partial\omega, \text{ where } Q(x) \geqslant 0.$$

By the maximum principle, $v \equiv 0$ in ω, $u_1 \equiv u_2$ in ω. From the maximum principle it also follows that $u_0 < 0$ in ω, since due to equation (7) $u_0(x)$ cannot attain in ω a positive maximum. The existence of u_0 can be proved using the Schauder fixpoint theorem (see [3]).

Let $u(x)$ be a solution of the problem (1),(2). From Lemma 1 it follows that $u(x) \leqslant 0$ in $S(X, \infty)$. From Lemma 3 we obtain that

$$u(x) = C_1 e^{\sqrt{\lambda_1}x_n}\varphi_1(x') + \mathcal{O}(1) \qquad (32)$$

with the constant $C_1 \leqslant 0$.

Suppose that $C_1 = 0$. Consider the function $v(x) = u_0(x') - u(x)$, where $u_0(x')$ is the solution of the problem (7),(8). We have

$$\Delta v = Q(x)v \text{ in } S(0, \infty), \qquad \text{where } Q(x) = \frac{e^{u_0} - e^u}{u_0 - u}.$$

Let us consider the function $v_1(x)$ such that

$$\Delta v_1 - Q(x)v_1 = 0 \text{ in } S(0,\infty), v_1 = 0 \text{ on } \sigma(0,\infty), \tag{33}$$

$$v_1 = v \text{ for } x_n = 0 \text{ and } \int_{S(0,\infty)} |\nabla v_1|^2 \, dx < \infty. \tag{34}$$

In order to construct the function v_1, we consider functions w_m such that

$$\Delta w_m - Q(x)w_m = 0 \text{ in } S(0,m), \tag{35}$$

$$w_m = 0 \text{ on } \sigma(0,m), \; w_m = v \text{ for } x_n = 0, \; w_m = 0 \text{ for } x_n = m. \tag{36}$$

In a standard way getting the energy estimate by multiplying (35) by $w_m e^{\alpha x_n}$ with $\alpha < \lambda_1$ and integrating it over $S(0,m)$ we obtain that

$$\int_{S(0,\infty)} |w_m|^2 e^{\alpha x_n} dx \; \leqslant \; C_2,$$

$$\int_{S(0,\infty)} |\nabla w_m|^2 e^{\alpha x_n} dx \; \leqslant \; C_3, \tag{37}$$

where C_2, C_3 are constants which do not depend on m. The estimates (37) are sufficient to allow us to pass to the limit in (35) and to prove, using in a standard way the E. De Giorgi estimate, that

$$\lim_{m \to \infty} w_m = v_1,$$

$$|v_1(x)| \leqslant C_4 e^{-\alpha x_n} \text{ in } S(0,\infty), \qquad \alpha, C_4 = const > 0. \tag{38}$$

Let us consider the function

$$V(x) = v(x) - v_1(x).$$

It is easy to see that V is a bounded function in $S(0,\infty)$ and we have

$$\Delta V - Q(x)V = 0 \text{ in } S(0,\infty),$$

$$V = 0 \text{ on } \sigma(0,\infty), V = 0 \text{ for } x_n = 0.$$

We set

$$W(x) = \epsilon Z_1(x') e^{\sqrt{\lambda_1} x_n}, \epsilon = const > 0,$$

where λ_1 is the first eigenvalue of the eigenvalue problem (5),(6) and $Z_1(x')$ is a corresponding eigenfunction, positive in ω. Consider

$$U_\pm = W \pm V \text{ in } S(0,M), \quad M = const < \infty.$$

We have

$$\Delta U_\pm - Q(x)U_\pm \leqslant 0 \text{ in } S(0,\infty),$$

$U_\pm \geqslant 0$ for $x_n = 0$, $U_\pm \geqslant 0$ for $x_n = M$, if M is sufficiently large, $U_\pm = 0$ on $\sigma(0,\infty)$.

The maximum principle, applied to $U_\pm$ in $S(0,M)$, gives us

$$U_\pm \geqslant 0 \text{ in } S(0,M).$$

Therefore

$$W \pm V \geqslant 0 \text{ in } S(0,\infty),$$
$$|V| \leqslant W = \epsilon Z_1(x')e^{\sqrt{\lambda_1}x_n}. \tag{39}$$

Since ϵ is arbitrary, it follows from (39) that $V \equiv 0$ in $S(0,\infty)$, $v_1 = v$, $u_0(x') - u(x) = v_1$. Hence $u(x) = u_0(x') - v_1(x)$ where the estimate (38) is valid for $v_1(x)$. Theorem 1 is proved.

The Neumann boundary condition. Let us consider the solution of equation (1) in $S(0,\infty)$ with the boundary condition

$$\frac{\partial u}{\partial \nu} = 0 \text{ on } \sigma(0,\infty), \tag{40}$$

where $\frac{\partial u}{\partial \nu}$ is the derivative of u in the exterior normal direction.

Theorem 2. *Let $u(x)$ be a solution of the problem (1),(40) in $S(0,\infty)$. Then*

$$u(x) \leqslant -2\ln(B + \frac{\sqrt{2}}{2}x_n) \text{ in } S(0,\infty), \tag{41}$$

where the constant B depends on $\max u(x',0)$.

Proof. We set

$$M = \max_\omega u(x',0).$$

Let $Y_\alpha(x_n)$ be a solution of the equation

$$Y_\alpha'' - e^{Y_\alpha} = 0 \tag{42}$$

with the initial conditions

$$
\begin{aligned}
Y_\alpha(0) &= M + 1, \\
Y_\alpha'(0) &= -\sqrt{2e^{M+1}} + \alpha, \qquad \alpha = const > 0.
\end{aligned} \tag{43}
$$

It is proved in Lemma 4 at the end of this lecture that the solution $Y_\alpha(x_n)$ exists such that
$$Y_\alpha(x_n) \to +\infty \text{ as } x_n \to X_\alpha < \infty.$$

Applying the maximum principle to $Y_\alpha(x_n) - u(x)$ in $S(0, X_\alpha)$ we obtain
$$u(x) \leqslant Y_\alpha(x_n) \text{ in } S(0, X_\alpha),$$

and $u(x) \leqslant Y_0(x_n) = -2\ln(B + \frac{\sqrt{2}}{2}x_n)$, where $B = e^{-\frac{(M+1)}{2}}$. The last inequality follows from the continuous dependence of the solutions of (42) on the initial data on a finite interval of the independent variable. The proof is complete.

Theorem 3. *Let $u(x)$ be a solution of the problem (1),(40) in $S(0,\infty)$. Then*
$$u(x) \geqslant C_1 x_n + C_2, \qquad C_1 = const < 0, C_2 = const. \tag{44}$$

Proof. According to Theorem 2 we have
$$\begin{aligned}
\Delta u &= e^u \\
&\leqslant e^{-2\ln(B + \frac{\sqrt{2}}{2}x_n)} \\
&= \frac{1}{(B + \frac{\sqrt{2}}{2}x_n)^2}, \qquad B = const > 0.
\end{aligned} \tag{45}$$

Let $v = -u$. Then we have
$$\Delta v = f(x) = -e^{-v}, \qquad v(x) \to +\infty \text{ as } x_n \to \infty, \tag{46}$$

$\frac{\partial u}{\partial \nu} = 0$ on $\sigma(0,\infty)$ and $0 \geqslant f(x) \geqslant -\frac{1}{(B + \frac{\sqrt{2}}{2}x_n)^2}$.

Let $w(x)$ be a solution of the linear equation
$$\Delta w = f(x) \text{ in } S(0,\infty) \tag{47}$$

such that $w = 0$ for $x_n = 0$, $\frac{\partial w}{\partial \nu} = 0$ on $\sigma(0,\infty)$ and
$$\int\limits_{S(0,\infty)} |\nabla w|^2 \, dx < \infty, \tag{48}$$

The solutions w can be obtained as a limit as $k \to \infty$ of the solution w_k of the boundary value problem
$$\Delta w_k = f(x) \text{ in } S(0,k), \tag{49}$$

$$w_k = 0 \text{ for } x_n = 0, w_k = 0 \text{ for } x_n = k \text{ and } \frac{\partial w_k}{\partial \nu} = 0 \text{ on } \sigma(0, k). \quad (50)$$

For functions w_k we have the energy inequality

$$\int_{S(0,k)} |\nabla w_k|^2 \, dx \leqslant \int_{S(0,k)} |w_k| \, |f(x)| dx$$

$$\leqslant \left(\int_{S(0,k)} \frac{|w_k|^2}{|x_n|^2} dx \right)^{1/2} \left(\int_{S(0,k)} |f|^2 |x_n|^2 dx \right)^{1/2}$$

$$\leqslant C_3 \left(\int_{S(0,k)} \frac{w_k^2}{x_n^2} dx \right)^{1/2}. \quad (51)$$

From (51) and the Hardy inequality

$$\int_{S(0,k)} \frac{|w_k|^2}{x_n^2} dx \leqslant C_4 \int_{S(0,k)} |\nabla w_k|^2 dx, \qquad C_4 = const,$$

where w_k is extended to $S(0, \infty)$ by defining $w_k = 0$ for $x_n > k$, we obtain

$$\int_{S(0,k)} |\nabla w_k|^2 dx + \int_{S(0,k)} \frac{w_k^2}{|x_n|^2} dx \leqslant C_5, \quad (52)$$

where the constant C_5 does not depend on k. From (52) it follows that a sub-sequence of $w_k(x)$ converges to $w(x)$ in $H^1(S(0, L))$ weakly as $k \to \infty$, $w(x)$ is a solution of (47),(48) and

$$\int_{S(0,\infty)} |\nabla w|^2 \, dx + \int_{S(0,\infty)} \frac{|w|^2}{x_n^2} dx \leqslant C_5. \quad (53)$$

Let $v_1 = v - w$. We then have

$$\Delta v_1 = 0 \text{ in } S(0, \infty), \frac{\partial v_1}{\partial \nu} = 0 \text{ on } \sigma(0, \infty).$$

For solutions of linear elliptic equations it is known that

$$|w(x)| \leqslant C_6 \left[\left(\int_{S(T-2,T+2)} w^2 dx \right)^{1/2} + \left(\int_{S(T-2,T+2)} |f|^2 dx \right)^{1/2} \right]$$

$$\text{for } x \in S(T-1, T+1). \quad (54)$$

From (53), (54) it follows that

$$|w(x)| \leqslant A_1 x_n + A_2, \qquad A_1, A_2 = const, \text{ in } S(0, \infty). \qquad (55)$$

Therefore,

$$v_2(x) \equiv v_1 + A_1 x_n + A_2,$$

for some constants A_1, A_2, is a positive function in $S(0, \infty)$, satisfying

$$\Delta v_2 = 0 \text{ in } S(0, \infty), \frac{\partial v_2}{\partial \nu} = 0 \text{ on } \sigma(0, \infty). \qquad (56)$$

Let

$$V(x_n) = \int\limits_\omega v_2(x', x_n) dx'.$$

From (56) it follows that

$$V'' = 0, V(x_n) = B_1 x_n + B_2.$$

Hence for any T there is a point x_T such that $x_T \in S(T - 1, T + 1)$ and

$$0 \leqslant v_2(x_T) \leqslant B_3 x_n + B_4. \qquad (57)$$

From (57) and the Harnack inequality it follows that

$$v_2(x) \leqslant B_5 x_n + B_6,$$
$$v_1(x) \leqslant B_7 x_n + B_8 \text{ and}$$
$$v(x) \leqslant B_9 x_n + B_{10}, \qquad B_j = const, \text{ in } S(0, \infty).$$

Since $u = -v$, we have

$$u(x) \geqslant -B_9 x_n - B_{10},$$

where $B_9 > 0$. The theorem is proved.

Theorem 4. *Let $u(x)$ be a solution of the problem (1),(40) and let*

$$\lim_{x_n \to \infty} \frac{u(x)}{\ln(\frac{\sqrt{2}}{2} x_n)} < -2. \qquad (58)$$

Then

$$u(x) \leqslant C_1 x_n + C_2, \qquad (59)$$

where $C_1, C_2 = const$ and $C_1 < 0$.

Proof. We set

$$u_1(x) = \frac{u(x)}{\ln\left(\frac{\sqrt{2}}{2}x_n\right)} + 2.$$

The function u_1 satisfies the equation

$$\Delta u_1 + \frac{2}{x_n \ln\left(\frac{\sqrt{2}}{2}x_n\right)} \frac{\partial u_1}{\partial x_n} - \frac{1}{x_n^2 \ln\left(\frac{\sqrt{2}}{2}x_n\right)} u_1$$

$$+ \frac{2}{x_n^2} \frac{\left(1 - e^{\left(\ln\left(\frac{\sqrt{2}}{2}x_n\right)u_1\right)}\right)}{u_1 \ln\left(\frac{\sqrt{2}}{2}x_n\right)} u_1$$

$$= 0.$$

In addition

$$\frac{\partial u_1}{\partial \nu} = 0 \text{ on } \sigma(0, \infty).$$

From Theorem 2 it follows that

$$u_1(x) \leqslant 0 \text{ in } S(0, \infty).$$

The function $u_2 = -u_1(x)$ satisfies the equation

$$\Delta u_2 + \frac{2}{x_n \ln\left(\frac{\sqrt{2}}{2}x_n\right)} \frac{\partial u_2}{\partial x_n} + Q_2(x)u_2 = 0,$$

where

$$Q_2(x) = -\frac{1}{x_n^2 \ln\left(\frac{\sqrt{2}}{2}x_n\right)} + \frac{2}{x_n^2} \frac{\left(1 - e^{-u_2\left(\ln\left(\frac{\sqrt{2}}{2}x_n\right)\right)}\right)}{u_2 \ln\left(\frac{\sqrt{2}}{2}x_n\right)}$$

and $u_2(x) \geqslant 0$ in $S(0, \infty)$.

It is easy to verify that $Q_2(x) \leqslant 0$ in $S(0, \infty)$. From the conditions of Theorem 4 it follows that

$$u_2(x^m) \geqslant \alpha_1 > 0$$

for a sequence $x^m = (x_1^m, \cdots, x_n^m)$ such that $x_n^m \to \infty$ as $m \to \infty$. From the Harnack inequality we get

$$u_2(x) \geqslant \alpha_2 > 0$$

for $x \in S(M, \infty)$ and sufficiently large M.

Assume that (59) is not valid. Then there exists a sequence of points $x^m = (x_1^m, \cdots, x_n^m)$ such that $x_n^m \to \infty$ and $\dfrac{u(x^m)}{x_n^m} \to 0$ as $x_n^m \to \infty$.

Therefore

$$\frac{u_2(x^m) \ln \left(\frac{\sqrt{2}}{2} x_n^m \right)}{x_n^m} \to 0 \text{ as } x_n^m \to \infty. \tag{60}$$

From (60) and the Harnack inequality it follows that

$$\frac{u_2(x) \ln \left(\frac{\sqrt{2}}{2} x_n \right)}{x_n} \to 0 \text{ in } S(x_n^m - 1, x_n^m + 1).$$

Consider the solution $Y_m(x_n)$ of the ordinary differential equation

$$L(Y) \equiv Y'' + \frac{2}{x_n \ln \left(\frac{\sqrt{2}}{2} x_n \right)} Y' - \frac{1}{x_n^2 \ln \left(\frac{\sqrt{2}}{2} x_n \right)} Y = 0 \tag{61}$$

subject to the boundary conditions

$$Y_m(2) = \max_{x' \in \omega} u_2(x', 2), Y_m(x_n^m) = \max_{x' \in \omega} u_2(x', x_n^m). \tag{62}$$

Such a solution of the problem (61),(62) exists and is unique. According to the maximum principle $Y_m > 0$. Equation (61) has a solution

$$Y(x_n) = \frac{x_n}{\ln \left(\frac{\sqrt{2}}{2} x_n \right)}.$$

From (60), (61), (62) and the maximum principle for the equation satisfied by $Y_m - \epsilon Y - \max_{x' \in \omega} u_2(x', 2)$ ($\epsilon = const > 0$), we get that

$$Y_m(x_n) \leqslant \epsilon Y(x_n) + \max_{x' \in \omega} u_2(x', 2) \tag{63}$$

for sufficiently large m.

From (63) it follows that there exists a limit for the sequence $Y_m(x_n)$ and this sequence converges to $Y_0(x_n)$ uniformly on any finite interval of x_n.

Since ϵ is arbitrarily small, it follows from (63) that

$$Y_0(x_n) = o(Y(x_n)) = o \left(\frac{x_n}{\ln \left(\frac{\sqrt{2}}{2} x_n \right)} \right). \tag{64}$$

The general solution of (61) has the form

$$Y(x_n) = C_3 \frac{x_n}{\ln\left(\frac{\sqrt{2}}{2}x_n\right)} + C_4 \frac{1}{\ln\left(\frac{\sqrt{2}}{2}x_n\right)}. \tag{65}$$

From (64), (65) it follows that

$$Y_0(x_n) = C_4 \frac{1}{\ln\left(\frac{\sqrt{2}}{2}x_n\right)} \tag{66}$$

so that $Y_0(x_n) \to 0$ as $x_n \to \infty$.

Let us apply the maximum principle to the equation which the function $u_2(x) - Y_m(x_n)$ satisfies. Since for $x_n = 2$ and for $x_n = x_n^m$ we have $u_2(x) - Y_m(x_n) \leqslant 0$, by the maximum principle,

$$u_2(x) \leqslant Y_m(x_n) \text{ in } S(2, x_n^m). \tag{67}$$

As $m \to \infty$, we get from (67) that

$$u_2(x) \leqslant Y_0(x_n) \text{ in } S(2, x_n^m).$$

But $Y_0(x_n) \to 0$ as $x_n \to \infty$ and this contradicts $u_2(x) \geqslant \alpha_2 > 0$ for sufficiently large x_n. Thus (59) is proved.

From Theorem 2–4 it follows that for any solution $u(x)$ of the problem (1),(40) we have either

$$u(x) = -2\ln x_n + o(\ln x_n) \text{ in } S(0, \infty), \text{ as } x_n \to \infty,$$

or

$$C_3 x_n + C_4 \leqslant u(x) \leqslant C_1 x_n + C_2,$$
$$C_j = const, C_1 < 0, \text{ in } S(0, \infty). \tag{68}$$

The following theorem gives a more precise result than (68).

Theorem 5. *If $u(x)$ is a solution of the problem (1),(40) and (68) is valid, then*

$$u(x) = Cx_n + o(x_n), C = const < 0, \text{ as } x_n \to \infty.$$

Proof. If for $u(x)$ the estimates (68) are valid, then $u(x)$ is a solution of the equation

$$\Delta u = f(x), \tag{69}$$

where $0 \leqslant f(x) \leqslant C_1 e^{-\delta x_n}$, $C_1, \delta = const > 0$.

Let us consider the equation

$$\Delta Z = F(x) \text{ in } S(-\infty, +\infty), \tag{70}$$

where $F(x) = f(x)$ for $x_n > 2$ and $F(x) = 0$ for $x_n < 2$, with the boundary condition

$$\frac{\partial Z}{\partial \nu} = 0 \text{ on } \sigma(-\infty, +\infty). \tag{71}$$

The problem (70),(71) has a solution such that

$$Z(x) = \mathcal{O}(e^{-\alpha x_n}) \text{ as } x_n \to \infty, \alpha = const > 0.$$

Then the function $Z_0(x) = u(x) - Z(x)$ satisfies the equation

$$\Delta Z_0 = 0 \text{ in } S(0, \infty), \frac{\partial Z_0}{\partial \nu} = 0 \text{ on } \sigma(0, \infty),$$

and from (68) it follows that

$$|Z_0(x)| \leqslant A_1 x_n + B_1.$$

It is known (see, for example, [4]) that

$$Z_0(x) = C x_n + o(x_n) \text{ as } x_n \to \infty, C = const < 0.$$

As a consequence of Theorems 2–5 we obtain the following.

Theorem 6. *Let $u(x)$ be a solution of the problem (1),(40) in $S(0, \infty)$ and $\frac{\partial u}{\partial \nu} = 0$ on $\sigma(0, \infty)$. Then either*

$$u(x) = -2 \ln x_n + o(\ln x_n) \text{ as } x_n \to \infty$$

or

$$u(x) = C x_n + o(x_n), C = const < 0, \text{ as } x_n \to \infty.$$

Now we shall prove a lemma for the ordinary differential equation

$$Y'' - e^Y = 0$$

which was used above to prove Theorems 1–6.

Lemma 4. *The equation (10) has a solution with the initial conditions*

$$Y(0) = Y_0, Y'(0) = Y_1$$

such that

$$Y(t) \to \infty \ as \ t \to A, A = const > 0, \ if \ Y_1 \geqslant 0; \tag{72}$$

$$Y(t) = -2\ln t + \mathcal{O}(1) \ as \ t \to \infty, \ if \ e^{Y_0} = \frac{1}{2}Y_1^2, Y_1 < 0; \tag{73}$$

$$Y(t) \to +\infty \ as \ t \to A = const > 0, \ if \ e^{Y_0} > \frac{1}{2}Y_1^2, Y_1 < 0; \tag{74}$$

$$Y(t) \to -\infty \ as \ t \to \infty, \ if \ e^{Y_0} < \frac{1}{2}Y_1^2, Y_1 < 0. \tag{75}$$

Proof. If $Y_1 \geqslant 0$, then $Y > Y_0$ and

$$(Y')^2 = 2\left(e^Y - e^{Y_0} + \frac{1}{2}Y_1^2\right). \tag{76}$$

From (76) it is easy to see that $Y(t) \to \infty$ as $t \to A$. In order to prove (73) we note that in this case

$$(Y')^2 = 2e^Y, Y' = -\sqrt{2}e^{Y/2}, Y(t) = -2\ln t + C_1, C_1 = const.$$

If we have $e^{Y_0} > \frac{1}{2}Y_1^2$ and $Y_1 < 0$, then

$$Y' = -\left(2\left(e^Y - e^{Y_0} + \frac{1}{2}Y_1^2\right)\right)^{1/2}. \tag{77}$$

From (77) it follows that either $Y(t) \to -\infty$ as $t \to \infty$ or $Y(t) \to A_1$, $A_1 = const < 0$, as $t \to \infty$. If $Y(t) \to -\infty$ as $t \to \infty$, then $e^Y - e^{Y_0} + \frac{1}{2}Y_1^2 < 0$ for sufficiently large t. It means that $Y(t) \to +\infty$ as $t \to A = const$, since we will have the case (74). If $Y(x) \to A_1$ as $t \to \infty$ then $\lim_{t\to\infty} Y''(t) = A_2$, $A_2 = const > 0$, as $t \to \infty$. This means that $Y'(t) \to \infty$ as $t \to \infty$ and $Y(t) \to +\infty$ as $t \to \infty$, which is impossible, since

$$Y' = -\left(2\left(e^Y - e^{Y_0} + \frac{1}{2}Y_1^2\right)\right)^{1/2} < 0.$$

Consider now the case $e^{Y_0} < \frac{1}{2}Y_1^2$. In this case as $t \to \infty$ either $Y(t) \to A$ or $Y(t) \to -\infty$. If $Y(t) \to A$, as $t \to \infty$, then

$$Y'(t) = -\left(2\left(e^Y - e^{Y_0} + \frac{1}{2}Y_1^2\right)\right)^{1/2} \to A_1 = const < 0,$$

as $t \to \infty$, and therefore $Y(t) \to -\infty$ as $t \to \infty$. Hence the case $Y(t) \to A = const$ as $t \to \infty$ cannot occur and we have $Y(t) \to -\infty$ as $t \to \infty$. The lemma is proved.

References

[1] O.A. Oleinik, On the equation $\Delta u + K(x)e^u = 0$, *Russian Math. Surveys* **33**, N 2, 1978.

[2] I. Kametaka, O.A. Oleinik, On asymptotic properties and necessary conditions to existence of solutions of nonlinear second order elliptic equations, *Mat. Sbornik* **107**, N 4, 1978, 572–600.

[3] R. Courant, Partial differential equations, Interscience Publishers, 1962.

[4] S.S. Lakhturov, On the asymptotic of solutions of the second boundary value problem in unbounded domain, *Uspekhi Mat. Nauk* **35**, N 4, 1980, 195–196.

2.5 On an approach to study asymptotics of solutions in cylindrical domains.

In this section we shall give a new approach to study the asymptotic behavior of solutions of differential equations in cylindrical domains.

We consider the elliptic equation of the form

$$\sum_{i,j=1}^{n-1} \frac{\partial}{\partial x_i}\left(a_{ij}(x)\frac{\partial u}{\partial x_j}\right) + \frac{\partial}{\partial x_n}\left(a_{nn}(x_n)\frac{\partial u}{\partial x_n}\right) + a_n(x_n)\frac{\partial u}{\partial x_n} - f(x_n, u) = 0 \tag{1}$$

in $S(0,\infty)$, where $a_{nn}(x_n)$, $a_{ij}(x)$, $a_n(x_n)$ are bounded measurable functions, $a_{nn} \geqslant a_0 = const > 0$, $a_n'(x_n) \to 0$ as $x_n \to \infty$, $a_{ij}(x) = a_{ji}(x)$, with the boundary condition

$$\frac{\partial u}{\partial \mu} = 0 \quad \text{on } \sigma(0,\infty). \tag{2}$$

We use here notations of section 1. We assume that the function $f(x_n, u)$ satisfies the conditions of Theorem 2, section 1: $f(x_n, u_1) > f(x_n, u_2)$ for $u_1 > u_2 \geqslant 0$, $f(x_n, u) \geqslant au^p$ for $u \geqslant 0$, $a = const > 0$, $p > 1$, the function $-f(x_n, -u)$ satisfies the same conditions as $f(x_n, u)$. In addition, we assume that $\frac{\partial f}{\partial u}$ is bounded for bounded u and $x \in S(0,\infty)$ and tends to zero as $u \to 0$ uniformly with respect to x_n. It is clear that $f(u) = |u|^{p-1}u$ satisfies these conditions.

We study the behavior of solutions of the problem (1),(2) as $x_n \to \infty$.

We assume that $\frac{\partial a_{ij}}{\partial x_i}$ are bounded in $S(0,\infty)$. According to Theorem 2, section 1 any solution of the problem (1),(2) tends to zero as $x_n \to \infty$ uniformly with respect to $x' = (x_1, \ldots, x_{n-1})$.

We set

$$U(x_n) = \frac{1}{|\omega|} \int\limits_{\omega(x_n)} u(x', x_n)\mathrm{d}x'$$

where $|\omega| = \int\limits_\omega \mathrm{d}x'$, $\omega(t) = \{x : x \in S(0,\infty), x_n = t\}$.

For simplicity we assume that $u \in C^1(\overline{S(0,\infty)}) \cap C^2(S(0,\infty))$.

Theorem 1. *Let $u(x)$ be a solution of the problem (1),(2). Then*

$$|u(x', x_n) - U(x_n)| \leqslant C_1 e^{-\alpha x_n}, \tag{3}$$

where C_1, α are positive constants and α does not depend on u.

Proof. Integrating equation (1) over $\omega(x_n)$, we get the equation for $U(x_n)$ of the form:

$$\frac{\partial}{\partial x_n}\left(a_{nn}(x_n)\frac{\partial U}{\partial x_n}\right) + a_n(x_n)\frac{\partial U}{\partial x_n} - \frac{1}{|\omega|}\int\limits_{\omega(x_n)} f(x_n, u)\mathrm{d}x' = 0. \quad (4)$$

For $W = U - u$ we have the equation

$$\sum_{i,j=1}^{n-1}\frac{\partial}{\partial x_i}\left(a_{ij}(x)\frac{\partial W}{\partial x_j}\right) + \frac{\partial}{\partial x_n}\left(a_{nn}(x_n)\frac{\partial W}{\partial x_n}\right) + a_n(x_n)\frac{\partial W}{\partial x_n}$$

$$+ f(x_n, u) - \frac{1}{|\omega|}\int\limits_{\omega(x_n)} f(x_n, u)\mathrm{d}x' = 0. \quad (5)$$

Let us transform the last two terms in (5). We have

$$f(x_n, u) - \frac{1}{|\omega|}\int\limits_{\omega(x_n)} f(x_n, u)\mathrm{d}x'$$

$$= f(x_n, u) - f(x_n, U) + f(x_n, U) - \frac{1}{|\omega|}\int\limits_{\omega(x_n)} f(x_n, u)\mathrm{d}x'$$

$$= f'_u W + \frac{1}{|\omega|}\int\limits_{\omega(x_n)} f'_u W \mathrm{d}x'$$

$$= \mathcal{F}_1 W + \int\limits_{\omega(x_n)} \mathcal{F}_2 W \mathrm{d}x',$$

where $\mathcal{F}_1 \to 0$, $\mathcal{F}_2 \to 0$ as $x_n \to \infty$, since $u \to 0$, $U \to 0$ as $x_n \to \infty$, and

$$f(x_n, U) - \frac{1}{|\omega|}\int\limits_{\omega(x_n)} f(x_n, u)\mathrm{d}x'$$

$$= \frac{1}{|\omega|}\int\limits_{\omega(x_n)} f(x_n, U)\mathrm{d}x' - \frac{1}{|\omega|}\int\limits_{\omega(x_n)} f(x_n, u)\mathrm{d}x'$$

$$= \frac{1}{|\omega|}\int\limits_{\omega(x_n)} f'_u W \mathrm{d}x'$$

$$= \int\limits_{\omega(x_n)} \mathcal{F}_2 W \mathrm{d}x'. \quad (6)$$

Thus, for U and W we have equations

$$\frac{\mathrm{d}}{\mathrm{d}x_n}\left(a_{nn}(x_n)\frac{\mathrm{d}U}{\mathrm{d}x_n}\right) + a_n(x_n)\frac{\mathrm{d}U}{\mathrm{d}x_n} - f(x_n, U) = -\int\limits_{\omega(x_n)} \mathcal{F}_2 W \mathrm{d}x', \quad (7)$$

$$\sum_{i,j=1}^{n-1} \frac{\partial}{\partial x_i}\left(a_{ij}(x)\frac{\partial W}{\partial x_j}\right) + \frac{\partial}{\partial x_n}\left(a_{nn}(x_n)\frac{\partial W}{\partial x_n}\right) + a_n(x_n)\frac{\partial W}{\partial x_n}$$

$$+\mathcal{F}_1 W + \int\limits_{\omega(x_n)} \mathcal{F}_2 W \mathrm{d}x' = 0 \qquad (8)$$

in $S(0, \infty)$ and the boundary condition

$$\frac{\partial W}{\partial \mu} = 0 \quad \text{on } \sigma(0, \infty).$$

Let $\varphi(t)$ be a function such that $\varphi(t) = 1$ for $t \leqslant 1$, $\varphi(t) = 0$ for $t \geqslant 2$, $0 \leqslant \varphi(t) \leqslant 1$, $\varphi \in C^\infty(\mathbf{R}^1)$. Multiplying the equation (8) by $W\varphi^2(\frac{x_n}{N})$, $N = const > 0$, and integrating it over $S(T, \infty)$, $T = const > 0$, we obtain

$$\lambda_1 \int\limits_{S(T,\infty)} |\nabla W|^2 \varphi^2 \mathrm{d}x$$

$$\leqslant \left|\int\limits_{S(T,\infty)} 2a_{nn}\frac{\partial W}{\partial x_n}W\varphi\frac{\partial \varphi}{\partial x_n}\mathrm{d}x\right| + \left|\int\limits_{S(T,\infty)} a_n\frac{\partial W}{\partial x_n}W\varphi^2\mathrm{d}x\right|$$

$$+\left|\int\limits_{\omega(T)} a_{nn}\frac{\partial W}{\partial x_n}W\mathrm{d}x'\right| + \left|\int\limits_{S(T,\infty)} \mathcal{F}_1 W^2\varphi^2\mathrm{d}x\right|$$

$$+\left|\int\limits_{S(T,\infty)} \left(\int\limits_{\omega(x_n)} \mathcal{F}_2 W\mathrm{d}x'\right) W\varphi^2\mathrm{d}x\right|, \qquad (9)$$

$\lambda_1 = const > 0$, $N > T$. Let $N \to \infty$. Then from (9) we have

$$\int\limits_{S(T,\infty)} |\nabla W|^2 \mathrm{d}x$$

$$\leqslant K_1 \left(\left|\int\limits_{S(T,\infty)} a_n W\frac{\partial W}{\partial x_n}\mathrm{d}x\right| + \left|\int\limits_{\omega(T)} a_{nn}\frac{\partial W}{\partial x_n}W\mathrm{d}x'\right|\right.$$

$$+\left|\int\limits_{S(T,\infty)}\mathcal{F}_1 W^2\mathrm{d}x\right|+\left|\int\limits_{S(T,\infty)}\left(\int\limits_{\omega(x_n)}\mathcal{F}_2 W\mathrm{d}x'\right)|W|\mathrm{d}x\right|\right),\quad(10)$$

since

$$\left|\int\limits_{S(T,\infty)}a_{nn}\frac{\partial W}{\partial x_n}W\varphi\frac{\partial\varphi}{\partial x_n}\mathrm{d}x\right|$$

$$\leqslant\epsilon\int\limits_{S(T,\infty)}|\nabla W|^2\varphi^2\mathrm{d}x+\frac{K_2}{\epsilon}\int\limits_{N<x_n<2N}W^2\left(\varphi'\left(\frac{x_n}{N}\right)\frac{1}{N}\right)^2\mathrm{d}x,$$

where ϵ is an arbitrary positive number, K_1, K_2 are constants and $W\to 0$ as $x_n\to\infty$ uniformly with respect to x', K_1, K_2 do not depend on T.

Taking into account the Poincaré inequality

$$\int\limits_{\omega(x_n)}W^2\mathrm{d}x'\leqslant C_1\int\limits_{\omega(x_n)}|\nabla W|^2\mathrm{d}x',$$

since

$$\int\limits_{\omega(x_n)}W\mathrm{d}x'=0,$$

we obtain

$$\left|\int\limits_{S(T,\infty)}a_n W\frac{\partial W}{\partial x_n}\mathrm{d}x\right|\leqslant\left|\int\limits_{S(T,\infty)}\frac{1}{2}\frac{\mathrm{d}a_n}{\mathrm{d}x_n}W^2\mathrm{d}x\right|+\left|\int\limits_{\omega(T)}a_n\frac{1}{2}W^2\mathrm{d}x'\right|,$$

$$\left|\int\limits_{\omega(T)}a_{nn}\frac{\partial W}{\partial x_n}W\mathrm{d}x'\right|\leqslant K_4\left(\int\limits_{\omega(T)}|\nabla W|^2\mathrm{d}x'\right)^{1/2}\left(\int\limits_{\omega(T)}|W|^2\mathrm{d}x'\right)^{1/2}$$

$$\leqslant K_5\int\limits_{\omega(T)}|\nabla W|^2\mathrm{d}x',$$

$$\left|\int\limits_{S(T,\infty)}\mathcal{F}_1 W^2\mathrm{d}x\right|\leqslant K_6\max_{S(T,\infty)}|\mathcal{F}_1|\int\limits_{S(T,\infty)}|\nabla W|^2\mathrm{d}x',$$

$$\left| \int\limits_{S(T,\infty)} \left(\int\limits_{\omega(x_n)} \mathcal{F}_2 |W| \mathrm{d}x' \right) |W| \mathrm{d}x \right|$$

$$\leqslant \left(\int\limits_{S(T,\infty)} |W|^2 \mathrm{d}x \right)^{1/2} \left(\int\limits_{S(T,\infty)} \left(\int\limits_{\omega(x_n)} \mathcal{F}_2 |W| \mathrm{d}x' \right)^2 \mathrm{d}x \right)^{1/2}$$

$$\leqslant K_7 \left(\int\limits_{S(T,\infty)} |\nabla W|^2 \mathrm{d}x \right)^{1/2}$$

$$\times \left(\int\limits_{S(T,\infty)} \left(\int\limits_{\omega(x_n)} |\mathcal{F}|^2 \mathrm{d}x' \int\limits_{\omega(x_n)} |\nabla W|^2 \mathrm{d}x' \right) \mathrm{d}x \right)^{1/2}$$

$$\leqslant K_8 \max_{S(T,\infty)} |\mathcal{F}_2| \int\limits_{S(T,\infty)} |\nabla W|^2 \mathrm{d}x.$$

From these inequalities and (10) we get

$$\int\limits_{S(T,\infty)} |\nabla W| \mathrm{d}x \leqslant K_9 \int\limits_{\omega(T)} |\nabla W|^2 \mathrm{d}x'$$

$$+ K_{10} \left(\max_{S(T,\infty)} |a'_n| + \max_{S(T,\infty)} |\mathcal{F}_1| + \max_{S(T,\infty)} |\mathcal{F}_2| \right) \int\limits_{S(T,\infty)} |\nabla W|^2 \mathrm{d}x.$$

Since $\mathcal{F}_1 \to 0$, $\mathcal{F}_2 \to 0$, $a'_n \to 0$ as $x_n \to \infty$, we obtain for T sufficiently large that

$$\int\limits_{S(T,\infty)} |\nabla W|^2 \mathrm{d}x \leqslant K_{11} \int\limits_{\omega(T)} |\nabla W|^2 \mathrm{d}x'. \tag{11}$$

We set

$$\int\limits_{S(T,\infty)} |\nabla W|^2 \mathrm{d}x \equiv J(T).$$

From (11) it follows that

$$J(T) \leqslant -K_{11} \frac{\mathrm{d}J}{\mathrm{d}T},$$

where the constant K_{11} does not depend on T. Integrating this inequality over the interval (T_0, T) we get for sufficiently large T that

$$\ln J(T) - \ln J(T_0) \leqslant -K_{12}(T - T_0), \quad J(T) \leqslant J(T_0)e^{-K_{12}(T-T_0)}.$$

Therefore

$$\int\limits_{S(T,\infty)} |\nabla W|^2 \mathrm{d}x \leqslant K_{13}e^{-K_{12}T}. \tag{12}$$

Using the Poincaré inequality for $W(x)$, we obtain from (12) that

$$\int\limits_{S(T,\infty)} |W|^2 \mathrm{d}x \leqslant K_{14}e^{-K_{12}T}, \tag{13}$$

where constants K_j do not depend on T.

Consider $W(x)$ as a solution of the equation (8) and consider the integral

$$\int\limits_{\omega(T)} \mathcal{F}_2 W \mathrm{d}x' = \phi(x_n)$$

as a given function, we can apply the De Giorgi estimate [1]–[3]. According to this estimate

$$\begin{aligned} \max_{S(T-1,T+1)} |W| &\leqslant K_{15}(\|W\|_{L^2(T-2,T+2)} + \|\phi\|_{L^q(T-2,T+2)}) \\ &\leqslant K_{16}\|W\|_{L^2(T-2,T+2)} \leqslant K_{17}e^{-\frac{1}{2}K_{12}T}, \quad q > \frac{n}{2}. \end{aligned} \tag{14}$$

The theorem is proved.

Thus, the problem on the asymptotic behavior of solutions $u(x)$ of the problem (1),(2) is reduced to study the asymptotic as $t \to \infty$ of solutions of the ordinary differential equation

$$\frac{\mathrm{d}}{\mathrm{d}t}\left(a_{nn}(t)\frac{\mathrm{d}U}{\mathrm{d}t}\right) + a_n(t)\frac{\mathrm{d}U}{\mathrm{d}t} - f(t,U) = \phi_1(t), \tag{15}$$

where the function ϕ_1 depends on t and

$$|\phi_1(t)| \leqslant K_{18}e^{-\alpha t}, \tag{16}$$

K_{18}, α are constants, K_{18}, $\alpha > 0$.

Using Theorem 1 and the equation (15) we can prove theorems on solutions of the problem (1),(2), similar to theorems sections 1–3.

Theorem 2. *Let $u(x)$ be a solution of the problem (1),(2), $\frac{\partial a_n}{\partial x_n} \to 0$ as $x_n \to \infty$, $a_n(x_n) + a'_{nn}(x_n) > a_0 = const > 0$. Then*

$$|u(x)| \leqslant M_1 e^{-\alpha x_n}, \quad M_1, \alpha = const > 0. \tag{17}$$

Proof. The function $Z(x_n) = e^{-\delta(x_n - T)}$, where $\delta = const > 0$, satisfies the inequality

$$a_{nn}Z'' + a'_{nn}Z' + a_n Z' - f(x_n, Z)$$
$$\leqslant (a_{nn}\delta^2 - \delta(a'_{nn} + a_n))e^{-\delta(x_n - T)}$$
$$\leqslant -M_2 e^{-\delta(x_n - T)}, \quad M_2 = const > 0, \tag{18}$$

if δ is sufficiently small. Let us take T so large that $U(x', T) \leqslant e^{-\delta(x_n - T)}$. For $Z - U = y$ we have the inequality

$$a_{nn}y + a'_{nn}y' + a_n y' - (f(x_n, Z) - f(x_n, U)) \leqslant -\phi_1(x_n) - M_2 e^{-\delta(x_n - T)}. \tag{19}$$

By virtue of (16) the right-hand side of (19) is less or equal to zero, if δ is sufficiently small. Since $\frac{\partial f}{\partial u} \geqslant 0$, from (19) and the maximum principle it follows that $y \geqslant 0$ for $x_n > T$, and

$$U(x) \leqslant e^{-\delta(x_n - T)}. \tag{20}$$

From (20) and Theorem 1 the estimate (17) follows.

Theorem 3. *Assume that $a_{nn} = 1$, $a_n = 0$, $f_u \leqslant C_0|u|^{p-1}$. Then the equation (15) has a unique solution $v(t)$ for which the estimate*

$$|v(t)| \leqslant C_1 e^{-\beta t}, \tag{21}$$

C_1, $\beta = const > 0$, *is valid. If for a solution v of (15) the estimate*

$$|v(t)| \leqslant C_2|t|^{\frac{2}{1-p} - \alpha_1} \tag{22}$$

is satisfied, where C_2, $\alpha_1 = const > 0$, then for $v(t)$ the estimate (21) is also valid.

Proof. A solution of the equation (15) which satisfies the inequality (21) can be constructed as a solution of the integral equation

$$y(t) = \int_t^\infty \int_{t_2}^\infty f(t_1, y)dt_1 dt_2 + \int_t^\infty \int_{t_2}^\infty \phi_1(t_1, y)dt_1 dt_2 \equiv K(y)$$

by the method of successive approximations, setting $y_0 \equiv 0$, $y_m = K(y_{m-1})$, $m = 1, 2, \ldots$.

In order to prove the uniqueness of such a solution consider $v_1 - v_2$, where v_1, v_2 are solutions of the equation (15), for which the estimate (21) is valid.

For $v_1 - v_2$ we have the equation

$$(v_1 - v_2)'' - Q(t)(v_1 - v_2) = 0 \quad \text{for } t > 0,$$

where $Q(t) \geqslant 0$ and

$$\int_1^\infty tQ(t)\mathrm{d}t < \infty.$$

It is known that in this case [4]

$$v_1 - v_2 = C_3(1 + o(1)) + C_4(t + o(t)), \quad C_3, C_4 = const, C_3^2 + C_4^2 \neq 0,$$

if $v_1 - v_2 \not\equiv 0$, $t \to \infty$. But $v_1 \to 0$ and $v_2 \to 0$ as $t \to \infty$. This contradiction proves that $v_1 \equiv v_2$. If v_1 satisfies (21) and v_2 satisfies (22), then in the same way we prove that $v_1 \equiv v_2$.

Theorem 4. *Let $u(x)$ be a solution of the problem (1),(2) and $a_{nn} = 1$, $a_n = 0$. If*

$$|u(x', x_n)| \leqslant C_5 |x_n|^{\frac{2}{1-p} - \alpha_1}, \quad C_5, \alpha_1 = const > 0,$$

then

$$|u(x', x_n)| \leqslant C_6 e^{-\alpha_2 x_n}$$

with some positive constants C_6, α_2.

Proof of this theorem follows from Theorem 1 and Theorem 3.

Using Theorem 1, one can get theorems on the asymptotic behavior of solutions of the problem (1),(2) in more general cases. The more general form of the equation (1) is considered in [5].

References

[1] E. De Giorgi, Sulla differenziabilitá e analiticita delle estremali degli integrali, *Mem. Accad. Sci. Torino*, 1957, 1–19.

[2] J. Moser, A new proof of the De Giorgi theorem concerning the regularity problem for elliptic differential equations, *Comm. Pure and Appl. Math.* **13**, 1960, 3, 457–468.

[3] D. Gilbarg, N.S. Trudinger, Elliptic partial differential equations of second order, 2nd edition, Springer Verlag, Berlin, 1983.

[4] R. Bellman, Stability theory of differential equations, Mc.Graw-Hill, New York, 1953.

[5] V.A. Kondratiev, O.A. Oleinik, On an approach to study asymptotics of solutions in cylindrical domains. *Russian Math. Survey*, **50**, N 4, 1995.

Chapter 3

On the asymptotics of solutions of nonlinear elliptic equations in a neighborhood of a conic point of the boundary

3.1 Asymptotic behavior of solutions of nonlinear elliptic equations near a conic point of the boundary with the Neumann boundary condition

The behavior of solutions of boundary-value problems for linear partial differential equations in a neighborhood of a nonregular point of the boundary has been studied in detail during the last few decades (see the survey [1] and Proceedings of the International Conference, which took place in Luminy (France) in May, 1993, and was dedicated to boundary-value problems in nonsmooth domains). For nonlinear equations not many results are known for this problem. It is interesting to note that in the nonlinear case new phenomena appear which do not occur in the case of linear equations.

Consider the equation

$$\sum_{i,j=1}^{n} \frac{\partial}{\partial x_i} \left(a_{ij}(x) \frac{\partial u}{\partial x_j} \right) - a_0(x)|u|^{p-1}u = 0 \tag{1}$$

in a bounded domain $\Omega \subset \mathbf{R}^n$, where $x = (x_1, \ldots, x_n)$, $p = const > 1$,

$$\lambda_1 |\xi|^2 \leqslant \sum_{i,j=1}^{m} a_{ij}(x)\xi_i\xi_j \leqslant \lambda_2 |\xi|^2, \quad \xi \in \mathbf{R}^n, x \in \Omega,$$

$a_0(x) \geqslant a_0 = const > 0$ and $a_{ij}(x) = a_{ji}(x)$, λ_1, $\lambda_2 = const > 0$. We assume that the domain Ω in a neighborhood of the origin $x = 0$ coincides with a cone K, whose vertex is $x = 0$. The domain K is called a cone with the vertex $x = 0$, if for any $x \in K$ the points λx also belong to K for any number $\lambda > 0$. Assume that the intersection K with a sphere $|x| = \delta$, $\delta > 0$, is a surface, satisfying a Lipschitz condition.

We set

$$
\begin{aligned}
K(a,b) &= \{x : x \in K, a < |x| < b\}, \\
\sigma(a,b) &= \partial K \cap \{x : a < |x| < b\},
\end{aligned}
$$

and suppose that Ω coincides with a cone K for $|x| < 1$. We consider a weak solution of equation (1) which satisfies the boundary condition

$$
\frac{\partial u}{\partial \mu} = 0 \text{ on } \sigma(0,1), \tag{2}
$$

where $\dfrac{\partial u}{\partial \mu} \equiv \displaystyle\sum_{i,j=1}^{n} a_{ij} \dfrac{\partial u}{\partial x_j} \nu_i$ and $\nu = (\nu_1, \cdots, \nu_n)$ is a unit outward normal vector to σ.

A weak solution of equation (1), satisfying the boundary condition (2), is a function $u(x)$ which belongs to $H^1(K(\epsilon,1)) \cap L^\infty(K(\epsilon,1))$ for any $0 < \epsilon < 1$ and satisfies the integral identity

$$
- \int_{K(0,1)} \sum_{i,j=1}^{n} a_{ij} \frac{\partial u}{\partial x_i} \frac{\partial \varphi}{\partial x_j} dx - \int_{K(0,1)} a_0(x)|u|^{p-1}u\varphi dx = 0 \tag{3}
$$

for any function $\varphi(x)$ which belongs to $H^1(K(0,1))$ and is equal to zero in $K(0,\delta)$ and $K(1-\delta,1)$ for some $\delta > 0$ (δ depends on φ).

First we prove some auxiliary results.

Lemma 1. *Let $u(x)$ be a solution of the equation (1) in the ball $|x-x^0| \leqslant \rho$, $\rho = const > 0$. Then in the ball $|x - x_0| \leqslant \rho/2$ the inequality*

$$
|u(x)| \leqslant C\rho^{\frac{2}{1-p}}
$$

holds, where the constant C does not depend on u.

This lemma is proved in the paper [2].

Lemma 2. *Let $u(x)$ be a solution of equation (1) in a domain G and $G \subset \{x : |x - x^0| < 1\}$, $\frac{\partial u}{\partial \mu} = 0$ on $\partial G \cap \{x : |x - x^0| < 1\}$. Suppose that the boundary ∂G of the domain G satisfies a Lipschitz condition. Then*

$$
|u(x)| \leqslant C_1
$$

in $G \cap \{x : |x - x^0| < 1/2\}$, *where the constant* C_1 *depends only on* λ_1, λ_2, a_0, n *and* G.

Proof. We cover $G \cap \{x : |x - x^0| < 1/2\}$ by balls of the fixed radius δ so small that the part of ∂G which belongs to a ball can be transformed by the Lipschitz transformation into a piece of plane. Then we extend the solution $u(x)$ outside G as an even function near ∂G and apply Lemma 1.

Lemma 3. *Let* $u(x)$ *be a solution of equation (1) in a domain* G *such that* $G \subset \{x : |x - x^0| < \rho\}$ *and satisfying the boundary condition* $\dfrac{\partial u}{\partial \mu} = 0$ *on* $\partial G \cap \{x : |x - x^0| < \rho\}$. *Then*

$$\int_{G \cap \{x : |x - x^0| < \rho/2\}} |\nabla u|^2 dx \leqslant C_2 \int_{G \cap \{x : |x - x^0| < \rho\}} \rho^{-2} |u|^2 dx. \qquad (4)$$

Proof. In order to prove (4) we put into the integral identity for $u(x)$ the function $\varphi = u\theta^2$, as a test function, where $\theta = 0$ for $|x - x^0| = \rho$, $\theta = 1$ for $|x - x^0| < \rho/2$, $|\nabla \varphi| \leqslant C_3 \rho^{-1}$.

Lemma 4. *Let* $u(x)$ *be a solution of equation (1) in* Ω *with the boundary condition (2) on* $\sigma(0,1)$. *Then*

$$|u(x)| \leqslant C_4 |x|^{\frac{2}{1-p}} \ in \ \Omega, \qquad (5)$$

$$\int_{K(\lambda/2, 2\lambda)} |\nabla u|^2 dx \leqslant C_5 \lambda^{\frac{2(1+p)}{1-p} + n}, \lambda < 1/4, \qquad (6)$$

C_4, C_5 *are constants,* C_5 *does not depend on* λ.

Proof. Consider $u(x)$ in $K(\lambda/4, 4\lambda)$ and change variables

$$x = \lambda y, \ u = v\lambda^{2/(1-p)}. \qquad (7)$$

The new function $v(y)$ is a solution of equation (1) in $K(1/4, 4)$, satisfying the boundary condition $\dfrac{\partial v}{\partial \mu} = 0$ on $\sigma(1/4, 4)$. Therefore for the function $v(y)$ one can apply Lemma 2 and the inequality (5) follows from Lemma 2 and (7). The inequality (6) follows from Lemma 3 and (5).

Theorem 1. *Let* $u(x)$ *be a weak solution of equation (1) in* $K(0,1)$ *satisfying the boundary condition (2) and let*

$$\frac{2}{p-1} \leqslant n - 2. \qquad (8)$$

Then

$$|u(x)| \leqslant C_6 \text{ in } \Omega, \tag{9}$$

and for $n > 2$

$$\int\limits_\Omega |\nabla u|^2 \mathrm{d}x \leqslant C_7, \quad C_6, C_7 = const. \tag{10}$$

Proof. We have from the inequality (5) and (8) that

$$|u(x)| \leqslant C_8 |x|^{2-n}, x \in K(0,1).$$

Suppose that $u(x)$ is not bounded in $K(0,1)$. Then we will show that there exists a positive function $u_1(x)$ such that $u_1(x)$ is unbounded in $K(0,1)$ and is a solution of the problem (1),(2) .

Let u_ρ be a solution of the equation (1) in $K(\rho,1)$ such that $u_\rho(x) = u_+(x)$ for $|x| = 1$ and for $|x| = \rho$, $\dfrac{\partial u}{\partial \mu} = 0$ on $\sigma(\rho,1)$, $0 < \rho < 1$, where $u_+(x)$ is, as usual, the positive part of $u(x)$. We can assume without loss of generality that $u_+(x)$ is an unbounded function in $K(0,1)$. The existence of the function $u_\rho(x)$ can be proved using the variational method. By the maximum principle (see [5],[9],[10])

$$u_\rho(x) \geqslant u(x) \text{ in } K(\rho,1).$$

From Lemma 2 we have that the sequence $u_\rho(x)$ as $\rho \to 0$ is uniformly bounded for any domain $K(\alpha,1)$, $\alpha = const > 0$. From well-known estimates of E. De Giorgi type it follows that $u_\rho(x)$ is compact in any domain $K(\alpha,1)$, $\alpha > 0$, in the sense of uniform convergence. They also converge weakly in $H^1(K(\alpha,1))$. The limit function for $u_\rho(x)$ as $\rho \to 0$ is the function $u_1(x)$ which we were looking for. Let us show that there exists a constant C_9 such that

$$u_1(x) \geqslant C_9 |x|^{2-n}. \tag{11}$$

We prove (11) by contradiction. Assume that (11) is not valid. Then there exists a sequence of points $x^m \to 0$ such that

$$u_1(x^m)|x^m|^{n-2} \to 0 \text{ as } m \to \infty. \tag{12}$$

We set $|x^m| = \lambda_m$. From (12) and the Harnack inequality it follows that

$$u_1(x)|x|^{n-2} \to 0, \text{ where } x \in K(\frac{\lambda_m}{2}, 2\lambda_m), \lambda_m \to 0. \tag{13}$$

From (13) and Lemma 3 we get that

$$\int_{K(\lambda_m/2, 2\lambda_m)} |\nabla u_1(x)|^2 dx = o\left(\lambda_m^{2-n}\right) \quad \text{as } \lambda_m \to 0. \tag{14}$$

Let us consider the identity (3) for $u_1(x)$ and set

$$\varphi(x) = \theta(x) \min(u_1(x), M),$$

where $\theta(x) = 0$ for $|x| < \frac{1}{2}\lambda_m$, $\theta(x) = 1$ for $\lambda_m < |x| < 3/4$, $\theta(x) = 0$ for $|x| \geqslant 1$, $\theta(x) \in C(\mathbf{R}^1)$, $0 \leqslant \theta \leqslant 1$. Let M be any constant such that $u_1(x) \leqslant M$ in $K(3/4, 1)$. For such a function φ the identity (3) gives us

$$\int_{K(0,1) \cap \{x: u_1(x) < M\}} \theta |\nabla u_1|^2 dx$$

$$\leqslant C_{10} \int_{K(\lambda_m/2, \lambda_m)} M|\theta||\nabla u_1| dx + C_{11}$$

$$\leqslant M\lambda_m^{\frac{n}{2}-1} \left(\int_{K(\lambda_m/2, \lambda_m)} |\nabla u_1|^2 dx \right)^{1/2} + C_{12}$$

$$\leqslant M o(1) + C_{12}, \qquad \text{as } m \to \infty.$$

Therefore

$$\int_{K(0,1)} |\nabla u_1|^2 dx \leqslant C_{13}, \tag{15}$$

where C_{13} is a constant.

Now let us take in the identity (3) $\varphi(x) = \theta(x) \max\{u_1(x), M\}$, where $\theta(x)$ is defined above. Then we have

$$\int_{K(0,1) \cap \{x: u_1(x) > M\}} \theta |\nabla u_1(x)|^2 dx$$

$$\leqslant C_{14} \lambda_m^{-1} \int_{K(\lambda_m/2, \lambda_m)} |u_1||\nabla u_1| dx$$

$$\leqslant C_{14} \left(\int_{K(\lambda_m/2, \lambda_m)} |\nabla u_1|^2 dx \right)^{1/2} \left(\int_{K(\lambda_m/2, \lambda_m)} \frac{|u_1|^2}{|x|^2} dx \right)^{1/2}$$

$$\to 0, \qquad \text{as } \lambda_m \to 0, \tag{16}$$

since $\int_{K(0,1)} \frac{|u_1|^2}{|x|^2} dx \leqslant C_{15}$ due to the Hardy inequality.

From (16) it follows that $u_1(x) \leqslant M$. This is a contradiction.

Therefore, any positive solution $u_1(x)$ of the problem (1),(2), which is unbounded, satisfies the inequality

$$u_1(x) \geqslant C_{16}|x|^{2-n} \text{ in } K(0,1). \tag{17}$$

From equation (1) we have

$$\sum_{i,j=1}^n \frac{\partial}{\partial x_i} \left(a_{ij} \frac{\partial u_1}{\partial x_j} \right)$$
$$= a_0(x)|u_1|^{p-1}u_1$$
$$\geqslant C_{17}|x|^{-n} \text{ in } K(0,1). \tag{18}$$

Let us take in the integral identity (3), $\varphi(x) = 0$ for $|x| = 1$, $\varphi(x) = 0$ for $|x| = \lambda$, $\varphi = 1$ for $2\lambda < |x| < 3/4$, $\varphi \geqslant 0$ in $K(0,1)$. Then we have, according to Lemma 3,

$$\int_{K(\lambda,3/4)} C_{17}|x|^{-n} dx \leqslant C_{18} \int_{K(\lambda,2\lambda)} |\nabla\varphi||\nabla u_1| dx$$
$$\leqslant C_{19}\lambda^{-1} \left(\int_{K(\lambda,2\lambda)} |\nabla u_1|^2 dx \right)^{\frac{1}{2}} \lambda^{n/2} + C_{20}$$
$$\leqslant C_{21}\lambda^{n-2} + C_{22} \leqslant C_{23},$$

where the constants C_{21}, C_{22}, C_{23} do not depend on λ. This contradiction proves that (9) is valid.

From (9) and Lemma 3 it follows that

$$\int_{K(\lambda/2,\lambda)} |\nabla u|^2 dx \leqslant C_{24}\lambda^{n-2} \text{ for } \lambda < 1. \tag{19}$$

Suppose that $n > 2$. Setting $\lambda = 2^{-k}$, $k = 0,1,\ldots$, and summing inequalities (19) over k from 0 to ∞ we obtain (10).

Theorem 2. *Let $u(x)$ be a solution of equation (1) in $K(0,1)$, satisfying the boundary condition (2) and*

$$\frac{2}{p-1} \leqslant n - 2. \tag{20}$$

Then there exists $u(0)$ *such that* $\lim\limits_{x \to 0} u(x) = u(0)$ *and*

$$|u(x) - u(0)| \leqslant C|x|^{\alpha} \qquad (21)$$

for some constants C *and* $\alpha; C, \alpha > 0$.

Proof. By Theorem 1 $u(x)$ is a solution of the linear equation

$$\sum_{i,j=1}^{n} \frac{\partial}{\partial x_i} \left(a_{ij}(x) \frac{\partial u}{\partial x_j} \right) = F(x), \qquad (22)$$

where $F(x)$ is the bounded function in $K(0, 1)$. Let $|F(x)| \leqslant M_1$. Let $v_1(x)$ be a solution of equation (22) in $K(0, 1)$ with the boundary condition

$$v_1 = 0 \text{ for } |x| = 1, \frac{\partial v_1}{\partial \mu} = 0 \text{ on } \sigma(0, 1)$$

and $v_1 \in H^1(K(0, 1))$. For v_1 we have the estimate

$$\int\limits_{K(0,1)} |\nabla v_1|^2 dx \leqslant C_1 \int\limits_{K(0,1)} M_1 |v_1| dx. \qquad (23)$$

It follows from (23) and the Friedrichs inequality that

$$\int\limits_{K(0,1)} |\nabla v_1|^2 dx + \int\limits_{K(0,1)} |v_1|^2 dx \leqslant C_2 M_1^2, \qquad (24)$$

where the constant C_2 does not depend on M_1. According to the E. De Giorgi theorem ([3]–[5]) we have

$$|v_1 x| \leqslant C_3 M_1 \text{ in } K(1/2, 1). \qquad (25)$$

For the function $v_\rho(x)$ which is a solution of (22) in $K(0, \rho)$ with the boundary conditions

$$v_\rho(x) = 0 \text{ for } |x| = \rho, \ \frac{\partial v_\rho}{\partial \mu} = 0 \text{ on } \sigma(0, \rho), \ v_\rho \in H^1(K(0, \rho)),$$

we have

$$|v_\rho(x)| \leqslant C_4 \rho^2 M_1 \text{ for } \rho/2 < |x| < \rho. \qquad (26)$$

The estimate (26) follows from the estimate (25) if we use the transformation

$$x = \rho y, v_1 = v_\rho. \qquad (27)$$

It is easy to see that for v_1 we have the estimate

$$|v_1(x)| \,\|_{|x|=2^{-m}}$$
$$\leqslant \max |v_1| \,\|_{|x|=2^{1-m}} + C_5 2^{-2m} M_1. \tag{28}$$

The inequality (28) is obtained by representing $v_1(x)$ as a sum of the solution w_1 of equation (22) in $K(0, 2^{-m+1})$ with $F(x) \equiv 0$ and with the boundary conditions

$$w_1 = v_1 \text{ for } |x| = 2^{1-m},$$
$$w_1 = 0 \text{ for } |x| = 2^{-m},$$
$$\frac{\partial w_1}{\partial \mu} = 0 \text{ for } \sigma(2^{-m}, 2^{-m+1})$$

and the solution w_2 of equation (22) in $K(0, 2^{-m+1})$ which we denote by v_{2-m+1}.

By iteration of inequality (28) we get that

$$|v_1(x)| \leqslant C_6 M_1 \text{ for } |x| = 2^{-m}. \tag{29}$$

Using the transformation (27) we obtain

$$|v_\rho(x)| \leqslant C_7 M_1 \rho^2 \text{ for } |x| < \rho, x \in K(0, \rho). \tag{30}$$

The function $w_\rho = u - v_\rho$ satisfies the equation

$$\sum_{i,j=1}^{n} \frac{\partial}{\partial x_i} \left(a_{ij}(x) \frac{\partial w_\rho}{\partial x_j} \right) = 0 \text{ in } K(0, \rho) \tag{31}$$

and the boundary condition $\frac{\partial w_\rho}{\partial \mu} = 0$ on $\sigma(0, \rho)$.

From the theory of linear second order elliptic equations it is known (see [3]–[5]) that if w is a weak solution of equation (31) in $K(0, \rho)$ with the boundary condition $\frac{\partial w}{\partial \mu} = 0$ on $\sigma(0, \rho)$, then

$$\text{osc } w|_{K(\rho/4, \rho/2)} \leqslant C_0 \text{osc } w|_{K(\rho/8, \rho)}, \quad C_0 < 1, \tag{32}$$

where

$$\text{osc } w|_\omega \equiv |\max_\omega w - \min_\omega w|.$$

Therefore we have, by iteration of (32) ,

$$\begin{aligned}
\text{osc } w_\rho|_{K(\rho 2^{-m-2}, \rho 2^{-m-1})} &\leqslant C_0^{m+1} \text{osc } w_\rho|_{K(\rho/8, \rho)} \\
&\leqslant C_0^{m+1} C_8 ,
\end{aligned} \tag{33}$$

$$|w_p| \leqslant \sup_{K(0,\rho)} |u| + \sup_{K(0,\rho)} |v_\rho| \leqslant C_9.$$

Let $\epsilon > 0$ be a given number. We choose ρ in such a way that

$$C_7 M_1 \rho^2 = \frac{\epsilon}{2}. \tag{34}$$

We now take m so large that

$$C_0^{m+1} C_g = \epsilon^{1/2}. \tag{35}$$

Summing up inequalities (33) over m from m to ∞, we have

$$\text{osc } w_\rho|_{K(0,\rho 2^{-m-1})} \leqslant C_{10}\epsilon^{1/2}, \qquad \rho = C_{11}\epsilon^{1/2}. \tag{36}$$

From the inequality (30) we have

$$|v_\rho(x)| \leqslant C_7 M_1 \rho^2 \text{ in } K(0,\rho),$$

and

$$\text{osc } v_s(x)|_{K(0,\rho 2^{-m-1})} \leqslant C_{12} \rho^2 2^{2(-m-1)}, \qquad s = \rho 2^{-m-1},$$

C_j are constants. Noting that the equality (35) defines

$$m = C_{13} + C_{14} \ln \epsilon,$$

we have from (34),(35) that

$$\rho^2 2^{2(-m-1)} \leqslant C_{15}\epsilon^l, \qquad l = const > 0, C_{15} = const > 0. \tag{37}$$

Hence

$$\begin{aligned}
&\text{osc } u|_{K(0,\rho 2^{-m-1})} \\
&\leqslant \text{osc } w_\rho|_{K(0,\rho 2^{-m-1})} + \text{osc } v_\rho|_{K(0,\rho 2^{-m-1})}.
\end{aligned} \tag{38}$$

From (36),(37),(38),(39) we get

$$\text{osc } u|_{K(0,C_{16}\epsilon^{1/2})} \leqslant C_{17}\epsilon^{1/2} + C_{18}\epsilon^l. \tag{39}$$

Therefore,

$$\text{osc } u|_{K(0,\delta)} \leqslant C_{18}\delta^\beta, \qquad \beta = const > 0,$$

for any arbitrarily small number δ. This means that (21) holds and the theorem is proved.

We note that condition (20) is essential for Theorem 2 to be valid. This leads to the following theorem.

Theorem 3. *Let $\frac{2}{p-1} > n - 2$, $p > 1$. Then there exists a solution of the problem (1),(2) of the form*

$$u(x) = C_p(1 + o(1))|x|^{\frac{2}{1-p}},$$

$$C_p = \left(\frac{2(1+p)}{(1-p)^2}\right)^{\frac{1}{p-1}} \quad as \ |x| \to 0.$$

3.2 Asymptotics of solutions of nonlinear elliptic equations near a conic point of the boundary with the Dirichlet boundary condition

The behavior of solutions of the equation

$$\sum_{i,j=1}^{n} \frac{\partial}{\partial x_i}\left(a_{ij}(x)\frac{\partial u}{\partial x_j}\right) - a_0(x)|u|^{p-1}u = 0 \text{ in } \Omega \qquad (1)$$

with the Dirichlet boundary condition in the neighborhood of the conic point of the boundary $\partial\Omega$ was studied in the paper [2] and [6]. We formulate here some of the results of this papers.

We use here the notations of section 1 and the assumptions made there about Ω and functions $a_{ij}(x)$, $a_0(x)$.

Let $\lambda(K)$ be the first eigenvalue of the problem

$$\Delta_\omega u + \lambda u = 0, \omega \in S_1, \qquad u = 0 \text{ on } \partial S_1,$$

where $S_1 = K \cap \{x : |x| = 1\}$, (r,ω) are polar coordinates with centre $x = 0$ and Δ_ω is the Beltrami operator. We set

$$\lambda_\pm(K) = \frac{1}{2}(2 - n \pm ((n-2)^2 + 4\lambda(K))^{1/2})$$

Theorem 1. *Let $u(x)$ be a weak solution of (1) with the boundary condition*

$$u = 0 \text{ on } \partial\Omega \cap Q_\beta, \qquad (2)$$

where β is a positive constant, $Q_\beta = \{x : |x| < \beta\}$, $u \in H^1(\Omega\backslash Q_\alpha) \cap L^\infty(\Omega\backslash Q_\alpha)$ for any $0 < \alpha < \alpha_0$, $p > 1$ and

$$\frac{2}{1-p} > \lambda_-(K).$$

Then

$$|u(x)| \leqslant C_\epsilon |x|^{\lambda_+(K)-\epsilon}, \qquad C_\epsilon, \epsilon = const > 0. \qquad (3)$$

The estimate (3) is valid for a solution $u \in H^1(\Omega)$ of the linear equation (1) with $a_0(x) \equiv 0$ and with condition (2). If

$$\frac{2}{1-p} < \lambda_-(K),$$

then there exist unbounded solutions of the problem (1),(2) in Ω.

Theorem 2. *Let $u(x)$ be a weak solution of (1),(2), $u \in H^1(\Omega)$,*

$$p < 1, \frac{2}{1-p} < \lambda_+(K).$$

Then $u \equiv 0$ in $\Omega \cap Q_\alpha$ for some $\alpha > 0$.

For singular solutions of nonlinear elliptic equations see [7], [8].

The results of Chapters 2 and 3 are obtained jointly with V.A. Kondratiev.

References

[1] V.A. Kondratiev, O.A. Oleinik, Boundary value problems for partial differential equations in nonsmooth domains, *Russian Math. Surveys* **38**, 1983, N 2.

[2] V.A. Kondratiev, E.M. Landis, On qualitative properties of solutions of a nonlinear second order equation,*Mat. Sbornik* **135**, 1988, N 3,346–360.

[3] E. De Giorgi, Sulla differenziabilità e l'analiticità delle estremali degli integrali, *Mem. Accad. Sci. Torino* 1957, 1–19.

[4] J. Moser, A new proof of the De Giorgi theorem concerning the regularity problem for elliptic differential equations, *Comm. Pure and Appl. Math.* **13**, 1960, N 3, 457–468.

[5] D. Gilbarg, N. S. Trudinger, Elliptic partial differential equations of second order, 2nd edition, Springer Verlag, Berlin, 1983.

[6] V.A. Kondratiev, On solutions of weakly nonlinear elliptic equations in a neighborhood of a conic point , *Differ.equations* **29**, 1993, N 2, 298–306.

[7] L. Veron, Singular solutions of some nonlinear elliptic equations, *Nonlinear Anal.* **5**, 1988, N 3, 225–242.

[8] H. Brezis, P.L. Lions, A note on isolated singularities for nonlinear elliptic equations, *Math. Anal. Appl.* **7**, A, 1981,263–266.

[9] G. Ficmera, Alcuni recenti sviluppi della theoria dei problemi al contorno per le equazioni alle derivate parziali lineari, *Atti del Conv. Internationale sulle equazioni alle derivate parziali*, 1954, Triest.

[10] M. Chicco, Principio di massimo per soluzioni di problemi al contorno misti per equazioni ellitiche di tipo variazionale, *Bull. Un. Math. Ital.* **4**, 1970, 3, 384–394.

Chapter 4

On some homogenization problems

The theory of homogenization of partial differential equations appeared about two decades ago and is a new branch of the theory of differential equations and mathematical physics. The theory of averaging (homogenization) had been developed much earlier for ordinary differential equations mainly in connection with problems of non linear mechanics.

In the field of partial differential equations the development of the homogenization theory was greatly stimulated by various problems arising in mechanics, physics, biology and modern technology, requiring asymptotic analysis based on the homogenization of differential operators.

The study of processes in strongly non-homogeneous media brings forth a large number of purely mathematical problems which are very important for applications. The theory of homogenization of differential operators and its applications form the subject of a vast literature. Two books on homogenization by the author of these lecture notes and her collaborators have been published recently: [1],[2] (see references there).

Here we consider some new problems of homogenization: the homogenization of second order elliptic equations in partially perforated domains with the Dirichlet boundary condition on holes and also with the Neumann boundary condition on holes. The same method can be used for the mixed boundary-value problem, (see [12],[14]). The problem with the Dirichlet condition on holes was considered in paper [3] where another approach was used. Homogenization problems in perforated domains were studied by many authors (see [1],[11] and the references therein). We consider here also corresponding vibration problems. In section 3 the asymptotics of solutions and eigenvalues of elliptic boundary value problems with rapidly alternating type of boundary conditions are studied.

4.1   On a homogenization problem for the Laplace operator in a partially perforated domain with the Dirichlet condition on holes

We use the following notation. Let Ω be a domain with a smooth boundary $\partial\Omega$. Assume that $\Omega \cap \{x\colon x_1 = 0\} \neq \emptyset$, $x = (x_1, \cdots, x_n)$, $\Omega \subset \mathbf{R}_x^n$. We set

$$\Omega^+ = \Omega \cap \{x\colon x_1 > 0\}, \qquad \Omega^- = \Omega \cap \{x\colon x_1 < 0\},$$
$$Q = \{\xi\colon 0 < \xi_j < 1, j = 1, \ldots, n\}, \qquad \xi = (\xi_1, \cdots, \xi_n),$$

Y_0 is a domain in $\mathbf{R}_\xi^n$ with a smooth boundary $\partial Y_0 = S_0$, $\overline{Y}_0 \subset Q$, $\overline{M}$ is the closure of a set M,

$$Y = Q \backslash \overline{Y}_0.$$

Let $\epsilon > 0$ be a small parameter. For any set X in $\mathbf{R}_x^n$, let

$$\epsilon X = \{x\colon \epsilon^{-1} x \in X\}, \qquad X + y = \{z\colon z = x + y, x \in X\}.$$

Let $g_\epsilon = \cup_m \epsilon(Y_0 + m)$, $g = \cup_m (Y_0 + m)$, $S = \cup_m (S_0 + m)$, where $m = (m_1, \ldots, m_n)$, $m_1 \geqslant 0$ and m_j $(j = 1, \ldots, n)$ are integers. We consider the domains

$$G_\epsilon = \mathbf{R}_x^n \backslash \overline{g_\epsilon}, \qquad \Omega_\epsilon^+ = \Omega^+ \cap G_\epsilon, \qquad \Omega_\epsilon = \Omega_\epsilon^+ \cup \Omega^- \cup \omega$$

where $\omega = \Omega \cap \{x\colon x_1 = 0\}$. Let

$$\Gamma_\epsilon = \partial\Omega_\epsilon \cap \partial\Omega, \qquad S_\epsilon = \partial\Omega_\epsilon \cap \Omega$$

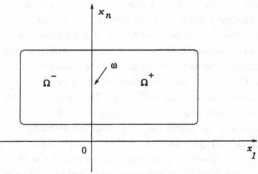

Figure 4.1: Domain Ω with a smooth boundary

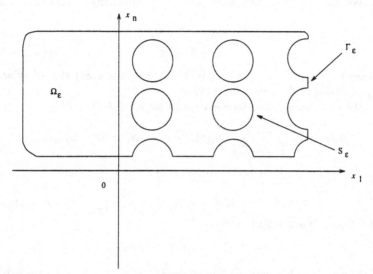

Figure 4.2: One cell

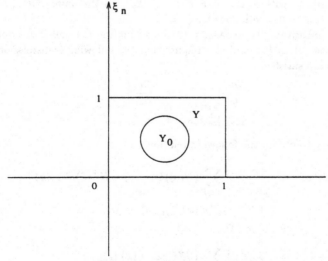

Figure 4.3: Partially perforated domain

We study the boundary-value problem in a partially perforated domain Ω_ϵ,

$$\Delta u_\epsilon = f \text{ in } \Omega_\epsilon, \tag{1}$$

$$u_\epsilon = 0 \text{ on } \partial\Omega_\epsilon, \tag{2}$$

where $f \in C^\infty(\overline{\Omega}^+)$ and $f \in C^\infty(\overline{\Omega}^-)$. Our aim is to study the behavior of the solutions of the problem (1),(2) as $\epsilon \to 0$.

We look for an asymptotic expansion for u_ϵ of the form

$$\begin{aligned}
\tilde{u}_\epsilon(x) &\approx \sum_{k=0}^\infty \epsilon^k \sum_{|i|=k} \big(N_i(\xi) D^i v_\epsilon(x) + N_i^+(\xi) D^i v_\epsilon(x) \,|_{x_1=+0} \\
&\quad + N_i^-(\xi) D^i v_\epsilon(x) \,|_{x_1=-0} \big),
\end{aligned} \tag{3}$$

where $\xi = \frac{x}{\epsilon}$, $i = (i_1, \ldots, i_k)$, $|i| = k$, $D^i = \dfrac{\partial^{|i|}}{\partial x_{i_1} \cdots \partial x_{i_k}}$, $i_j = 1, \ldots, n$, for $|i| > 0$ and $D^i v \equiv v$ for $|i| = 0$,

$$v_\epsilon(x) = v_0(x) + \epsilon v_1(x) + \epsilon^2 v_2(x) + \cdots, \tag{4}$$

$N_i(\xi)$ is a smooth 1-periodic function in $\{\xi : \xi_1 \leqslant 0\}$ and also a smooth 1-periodic function in $\{\xi : \xi_1 \geqslant 0\}$, $N_i^+(\xi)$, $N_i^-(\xi)$ are functions of a boundary layer type which tend to zero as $|\xi_1| \to \infty$. They are defined in $\mathbf{R}^n_\xi \backslash g$ and are 1-periodic with respect to $\hat{\xi} = (\xi_2, \cdots, \xi_n)$. Furthermore, $N_i(\xi)$, $N_i^+(\xi)$, $N_i^-(\xi)$ can be discontinuous for $\xi_1 = 0$.

We substitute the expansion (3) into equation (1) and into boundary condition (2) on $\partial\Omega_\epsilon$ and also require that the following transmission conditions are satisfied on ω:

$$\begin{aligned}
\frac{\partial \tilde{u}_\epsilon}{\partial x_1}\bigg|_{x_1=+0} &= \frac{\partial \tilde{u}_\epsilon}{\partial x_1}\bigg|_{x_1=-0}, \\
\tilde{u}_\epsilon\,|_{x_1=+0} &= \tilde{u}_\epsilon\,|_{x_1=-0}.
\end{aligned}$$

It is easy to obtain the following relations:

$$\begin{aligned}
\Delta \tilde{u}_\epsilon(x) &\approx \sum_{k=0}^\infty \epsilon^{k-2} \sum_{|i|=k} \big(H_i(\xi) D^i v_\epsilon(x) + H_i^+(\xi) D^i v_\epsilon(x) \,|_{x_1=+0} \\
&\quad + H_i^-(\xi) D^i v_\epsilon(x) \,|_{x_1=-0} \big) \text{ in } \Omega_\epsilon, \tag{5}
\end{aligned}$$

$$\tilde{u}_\epsilon \approx 0 \text{ on } \partial\Omega_\epsilon, \tag{6}$$

$$\left[\frac{\partial \tilde{u}_\epsilon}{\partial x_1}\right] \approx \sum_{k=0}^\infty \epsilon^{k-1} \sum_{|i|=k} \Big(J_i^+(\hat{\xi}) D^i v_\epsilon(x) \,|_{x_1=+0}$$

$$+J_i^-(\hat{\xi})D^i v_\epsilon(x)\,|_{x_1=-0}\Big)\ \text{on}\ \omega, \tag{7}$$

$$[\tilde{u}_\epsilon]\ =\ \sum_{k=0}^{\infty}\epsilon^k\sum_{|i|=k}\Big(G_i^+(\hat{\xi})D^i v_\epsilon(x)\,|_{x_1=+0}$$

$$+G_i^-(\hat{\xi})D^i v_\epsilon(x)\,|_{x_1=-0}\Big)\ \text{on}\ \omega. \tag{8}$$

Here and in what follows, we use the notation

$$[w]=w\,|_{x_1=+0}-w\,|_{x_1=-0}\,.$$

For the construction of the asymptotic expansion for u_ϵ we choose functions $N_i(\xi)$, $N_i^+(\xi)$, $N_i^-(\xi)$ in the following way:

$$
\begin{align}
H_0(\xi)\ &=\ 0,\ \ H_{i_1}(\xi)=0\ \text{for}\ i_1=1,\dots,n, \tag{9}\\
H_i(\xi)\ &=\ h_i^+\ \text{for any}\ i\ \text{with}\ |i|\geqslant 2,\ \text{for}\ \xi_1>0, \tag{10}\\
H_i(\xi)\ &=\ h_i^-\ \text{for any}\ i\ \text{with}\ |i|\geqslant 2,\ \text{for}\ \xi_1<0, \tag{11}\\
H_i^-(\xi)\ &=\ 0, H_i^+(\xi)=0\ \text{for any}\ i\ \text{and}\ \xi\in \mathbf{R}_\xi^n\backslash g, \tag{12}\\
J_i^+(\hat{\xi})\ &=\ j_i^+\ \text{for any}\ i, \tag{13}\\
J_i^-(\hat{\xi})\ &=\ j_i^-\ \text{for any}\ i, \tag{14}\\
G_i^+(\hat{\xi})\ &=\ g_i^+\ \text{for any}\ i, \tag{15}\\
G_i^-(\hat{\xi})\ &=\ g_i^-\ \text{for any}\ i, \tag{16}
\end{align}
$$

where h_i^+, h_i^-, j_i^+, j_i^-, g_i^+, g_i^- are some constants which we choose later. Calculations show that equations (9)–(16) lead to the following boundary-value problems for the functions $N_i(\xi)$, $N_i^+(\xi)$, $N_i^-(\xi)$.

For function $N_i(\xi)$ for $\xi_1>0$ we have

$$
\left\{
\begin{align}
&\Delta N_0=0\ \text{in}\ Y, \tag{17}\\
&N_0=0\ \text{on}\ S_0, \tag{18}\\
&N_0(\xi)\ \text{is 1-periodic in}\ \xi; \tag{19}
\end{align}
\right.
$$

$$
\left\{
\begin{align}
&\Delta N_{i_1}+2D^{i_1}N_0=0\ \text{in}\ Y, i_1=1,\dots,n, \tag{20}\\
&N_{i_1}=0\ \text{on}\ S_0, \tag{21}\\
&N_{i_1}\ \text{is 1-periodic in}\ \xi\ ; \tag{22}
\end{align}
\right.
$$

$$\begin{cases} \Delta N_{i_1 i_2} + 2D^{i_2} N_{i_1} + N_0 \delta_{i_1 i_2} = h_{i_1 i_2}^+ \text{ in } Y, & (23) \\ N_{i_1 i_2} = 0 \text{ on } S_0, & (24) \\ N_{i_1 i_2} \text{ is 1-periodic in } \xi; i_1, i_2 = 1, \dots, n. & (25) \end{cases}$$

For any $i = (i_1, \dots, i_k)$, $|i| > 2$ we have the boundary-value problem

$$\begin{cases} \Delta N_i + 2D^{i_k} N_{i_1 \dots i_{k-1}} + \delta_{i_{k-1} i_k} N_{i_1 \dots i_{k-2}} = h_i^+ \text{ in } Y, & (26) \\ N_i = 0 \text{ on } S_0, & (27) \\ N_i \text{ is 1-periodic in } \xi \text{ for any } i \text{ and } i_j = 1, \dots, n; & (28) \end{cases}$$

δ_{ks} is the Kronecker symbol: $\delta_{ks} = 1$ for $k = s$ and $\delta_{ks} = 0$ for $k \neq s$.

For $\xi_1 < 0$ for functions N_i we have equations of the form (17), (20), (23), (26) in Q, where the constants h_i^+ are replaced by h_i^-.

For functions N_i^+ and N_i^- which are of the boundary layer type and are defined in $\mathbf{R}_\xi^n \backslash g$ we have

$$\begin{cases} \Delta N_0^\pm = 0 \text{ in } \mathbf{R}_\xi^n \backslash g, & (29) \\ N_0^\pm = 0 \text{ on } S, & (30) \\ [N_0^\pm] = \mp N_0 \left.\right|_{\xi_1 = \pm 0} + g_0^\pm, & (31) \\ \left[\dfrac{\partial N_0^\pm}{\partial \xi_1}\right] = \mp \dfrac{\partial N_0}{\partial \xi_1}\bigg|_{\xi_1 = \pm 0} + j_0^\pm, & (32) \\ |N_0^\pm(\xi)| \to 0 \text{ as } |\xi_1| \to \infty, N_0^\pm \text{ is 1-periodic in } \hat{\xi}, \\ \qquad \hat{\xi} = (\xi_2, \dots, \xi_n); \end{cases}$$

$$\begin{cases} \Delta N_{i_1}^\pm + 2(1 - \delta_{i_1 1}) D^{i_1} N_0^\pm = 0 \text{ in } \mathbf{R}_\xi^n \backslash g, & (33) \\ N_{i_1}^\pm = 0 \text{ on } S, & (34) \\ [N_{i_1}^\pm] = \mp N_{i_1} \left.\right|_{\xi_1 = \pm 0} + g_{i_1}^\pm, & (35) \\ \left[\dfrac{\partial N_{i_1}^\pm}{\partial \xi_1}\right] = \mp \left(\dfrac{\partial N_{i_1}}{\partial \xi_1} + \delta_{i_1 1} N_0\right)_{\xi_1 = \pm 0} + j_{i_1}^\pm, & (36) \\ |N_{i_1}^\pm(\xi)| \to 0 \text{ as } |\xi_1| \to \infty, N_{i_1}^\pm \text{ is 1-periodic in } \hat{\xi}, & (37) \\ \qquad i_1 = 1, \dots, n; \end{cases}$$

$$\left\{ \begin{array}{l} \Delta N_{i_1 i_2}^{\pm} + (1 - \delta_{i_1 1})\left(2D^{i_1} N_{i_2}^{\pm} + N_0^{\pm}\delta_{i_1 i_2}\right) = 0 \\ \qquad \text{in } \mathbf{R}^n \backslash g, \\ N_{i_1 i_2}^{\pm} = 0 \text{ on } S, \\ \left[N_{i_1 i_2}^{\pm}\right] = \mp N_{i_1 i_2} \mid_{\xi_1 = \pm 0} + g_{i_1 i_2}^{\pm}, \\ \left[\dfrac{\partial N_{i_1 i_2}^{\pm}}{\partial \xi_1}\right] = \mp \left(\dfrac{\partial N_{i_1 i_2}}{\partial \xi_1} + \delta_{i_1 1} N_{i_2}\right) \mid_{\xi_1 = \pm 0} + j_{i_1 i_2}^{\pm}, \\ \left|N_{i_1 i_2}^{\pm}(\xi)\right| \to 0 \text{ as } |\xi_1| \to \infty, \\ N_{i_1 i_2}^{\pm} \text{ is 1-periodic in } \hat{\xi}, i_1, i_2 = 1, \ldots, n. \end{array} \right.$$

$$(38)$$
$$(39)$$
$$(40)$$
$$(41)$$
$$(42)$$
$$(43)$$

For any $i = (i_1, \ldots, i_k)$ we have

$$\left\{ \begin{array}{l} \Delta N_i^{\pm} + (1 - \delta_{i_1 1})\left(2D^{i_1} N_{i_2 \cdots i_k}^{\pm} + N_{i_3 \cdots i_k}^{\pm}\delta_{i_1 i_2}\right) \\ = 0 \qquad \text{in } \mathbf{R}^n \backslash g, \\ N_i^{\pm} = 0 \text{ on } S, \\ \left[N_i^{\pm}\right] = \mp N_i \mid_{\xi_1 = \pm 0} + g_i^{\pm}, \\ \left[\dfrac{\partial N_i^{\pm}}{\partial \xi_1}\right] = \mp \left(\dfrac{\partial N_i}{\partial \xi_1} + \delta_{i_1 1} N_{i_2 \cdots i_k}\right)\Bigg|_{\xi_1 = \pm 0} + j_i^{\pm}, \\ \left|N_i^{\pm}(\xi)\right| \to 0 \text{ as } |\xi_1| \to \infty, N_i^{\pm} \text{ is 1-periodic in } \hat{\xi}. \end{array} \right.$$

$$(44)$$
$$(45)$$
$$(46)$$
$$(47)$$

The boundary-value problems for the functions $N_i^{\pm}(\xi)$ have solutions only for some constants $g_i^{\pm}$, $j_i^{\pm}$. The boundary-value problems for $N_i(\xi)$ have solutions for $\xi_1 > 0$ and any h_i^+. For $\xi_1 < 0$ the problems for $N_i(\xi)$ have solutions only for some constants h_i^- which we will choose later, [6].

In order to prove the existence of solutions of problems for $N_i^{\pm}$ we will use the following theorem. Let us set

$$\Pi = (\mathbf{R}_{\xi}^n \backslash g) \cap \{\xi : 0 < \xi_j < 1, j = 2, \ldots, n\}, \text{ and}$$
$$\gamma = \Pi \cap \{\xi : \xi_1 = 0\}.$$

Theorem 1. *For any constant J there exists a constant q such that the problem*

$$\Delta N = -f_0 - \sum_{j=1}^n \frac{\partial f_j}{\partial x_j} \text{ in } \mathbf{R}_{\xi}^n \backslash g,$$

$$(48)$$

$$N = 0 \ on \ S, \tag{49}$$

$$[N] = V(\hat{\xi}) + q \ on \ \gamma, \tag{50}$$

$$\left[\frac{\partial N}{\partial \xi_1}\right] = W(\hat{\xi}) + J \ on \ \gamma, \tag{51}$$

$$|N(\xi)| \to 0 \ as \ |\xi_1| \to 0, N(\xi) \ is \ 1\text{-periodic in } \hat{\xi}, \tag{52}$$

has a solution $N(\xi)$ and

$$|N(\xi)| \leqslant C_2 e^{-\alpha_2 |\xi_1|}, \ C_2, \alpha_2 = const > 0, \tag{53}$$

where $|f_j(\xi)| \leqslant C_1 e^{-\alpha_1 |\xi_1|}$, f_j is 1-periodic in $\hat{\xi}$, $f_j \in H^1_{loc}(\Pi)$, $j = 0, 1, \ldots, n$, C_1, $\alpha_1 = const > 0$, $W(\hat{\xi})$, $V(\hat{\xi})$ belong to $H^1(\gamma)$.

Proof. Let $P_M = \left(\mathbf{R}^n_\xi \backslash g\right) \cap \{\xi : |\xi_1| < M\}$, $M = const > 0$, $\Pi_M = \Pi \cap \{\xi : |\xi_1| < M\}$. Consider the following boundary-value problem:

$$\Delta U_M = -f_0 - \sum_{j=1}^n \frac{\partial f_j}{\partial x_j} \ in \ P_M, \tag{54}$$

$$U_M = 0 \ on \ S_M, S_M = S \cap P_M, \tag{55}$$

$$[U_M] = V(\hat{\xi}) \ on \ \gamma, \tag{56}$$

$$\left[\frac{\partial U_M}{\partial \xi_1}\right] = W(\hat{\xi}) + J \ on \ \gamma, \tag{57}$$

$$U_M = 0 \ for \ \xi_1 = -M, U_M = 0 \ for \ \xi_1 = M, \tag{58}$$

$$and \ U_M \ is \ 1\text{-periodic in } \hat{\xi}. \tag{59}$$

We set

$$W_M(\xi) = U_M(\xi) - V(\hat{\xi})\psi(\xi_1)$$

where $\psi(\xi_1)$ is such that $\psi(\xi_1) = 1$ for $0 \leqslant \xi_1 \leqslant \delta$, $\psi(\xi_1) = 0$ for $\xi_1 < 0$ and for $\xi_1 \geqslant 2\delta$, $0 \leqslant \psi(\xi_1) \leqslant 1$, $\psi \in C^\infty(\mathbf{R}^1_+)$, where δ is a sufficiently small constant, $\mathbf{R}^1_+ = \{\xi_1 : \xi_1 \geqslant 0\}$, $\psi = 0$ on S.

We define the space $H^1(G, \Gamma)$ as a completion of the class of functions $u \in C^\infty(\overline{G})$ which are equal to zero in a neighborhood of Γ, by the norm

$$\|u\| = \left(\int\limits_G \left(|\nabla u|^2 + u^2\right) dx\right)^{1/2}.$$

We set $\Gamma_M = S_M \cup \{\xi : \xi_1 = -M, \xi_1 = M\}$. The function $U_M \in H^1(\Pi_M)$ is called a weak solution of the problem (54)–(59), if the function $W_M(\xi)$ for any $\varphi \in H^1(\Pi_M, \Gamma_M)$ which is 1-periodic in $\hat{\xi}$ satisfies the integral identity

$$
\begin{aligned}
-\int\limits_{\Pi_M} \nabla W_M \nabla \varphi \mathrm{d}\xi \;=\; & -\int\limits_{\Pi_M} f_0 \varphi \mathrm{d}\xi + \int\limits_{\Pi_M} \sum_{j=1}^n f_j \frac{\partial \varphi}{\partial \xi_j} \mathrm{d}\xi \\
& + \int\limits_{\Pi_M^+} \nabla(\psi V) \nabla \varphi \mathrm{d}\xi + \int\limits_\gamma [f_1] \varphi \mathrm{d}\hat{\xi} \\
& + \int\limits_\gamma (W + J) \varphi \mathrm{d}\hat{\xi}, \qquad (60)
\end{aligned}
$$

where $\Pi_M^+ = \Pi_M \cap \{\xi : \xi_1 > 0\}$, $\nabla u = grad\; u \equiv \left(\frac{\partial u}{\partial \xi_1}, \ldots, \frac{\partial u}{\partial \xi_n}\right)$. Since W_M is equal to zero on Γ_M, the Friedrichs inequality is valid:

$$
\int\limits_{\Pi_M} (W_M)^2 \mathrm{d}\xi \leqslant C_3 \int\limits_{\Pi_M} |\nabla W_M|^2 \mathrm{d}\xi, \qquad (61)
$$

where the constant C_3 depends on M.

For the proof of the existence of a weak solution of the problem (54)–(59) it is sufficient in view of the Lax–Milgram theorem [4] to prove that the right hand side of (60) is a linear bounded functional of φ in $H(\Pi_M, \Gamma_M)$. We have

$$
\begin{aligned}
& \left| -\int\limits_{\Pi_M} f_0 \varphi \mathrm{d}\xi + \int\limits_{\Pi_M} \sum_{j=1}^n f_j \frac{\partial \varphi}{\partial \xi_j} \mathrm{d}\xi + \int\limits_{\Pi_M^+} \nabla(\Psi V) \nabla \varphi \mathrm{d}\xi + \int\limits_\gamma [f_1] \varphi \mathrm{d}\hat{\xi} \right. \\
& \left. + \int\limits_\gamma (W + J) \varphi \mathrm{d}\hat{\xi} \right| \\
& \leqslant C_4 \left(\int\limits_{\Pi_M} |\nabla \varphi|^2 \mathrm{d}\xi \right)^{1/2},
\end{aligned}
$$

since from (61) and the imbedding theorem

$$
\int\limits_\gamma |\varphi|^2 \mathrm{d}\hat{\xi} \leqslant C_5 \left(\int\limits_{\Pi_M} \varphi^2 \mathrm{d}\xi + \int\limits_{\Pi_M} |\nabla \varphi|^2 \mathrm{d}\xi \right) \leqslant C_6 \int\limits_{\Pi_M} |\nabla \varphi|^2 \mathrm{d}\xi, \qquad (62)
$$

where the constants C_4, C_5, C_6 do not depend on φ. Now we shall prove that

$$\int_{\Pi_M} |\nabla W_M|^2 \, d\xi \leqslant K_1, \tag{63}$$

where the constant K_1 does not depend on M.

Let us define W_M in $\{\xi : |\xi_1| < M\}$, setting

$$W_M = 0 \text{ in } \{\xi : |\xi_1| < M\} \backslash P_M.$$

We set

$$Q_M = \{\xi : 0 < \xi_j < 1, j = 2, \ldots, n, |\xi_1| < M\}.$$

We note that for W_M the inequality (61) is valid in Q_M. It is easy to see that

$$\int_{Q_M} W_M^2 \left(|\xi|^3 + 1\right)^{-1} d\xi$$

$$\leqslant 4 \int_\gamma W_M^2 d\hat{\xi} \int_0^M \left(|\xi_1|^3 + 1\right)^{-1} d\xi_1$$

$$+ 4 \int_{Q_M} \left|\frac{\partial W_M}{\partial \xi_1}\right|^2 d\xi \int_0^M |\xi_1| \left(|\xi_1|^3 + 1\right)^{-1} d\xi_1. \tag{64}$$

Therefore, using (62) for Π_1 and applying the Friedrichs inequality for Π_1 we have

$$\int_{Q_M} W_M^2 \left(|\xi_1|^3 + 1\right)^{-1} d\xi \leqslant C_7 \int_{Q_M} |\nabla W_M|^2 d\xi \tag{65}$$

with a constant C_7 independent of M. In order to prove (64) we consider the inequality

$$W_M^2(\xi) \leqslant 2W_M^2(\xi^0) + 2|\xi_1| \int_0^{\xi_1} \left(\frac{\partial W_M}{\partial \xi_1}\right)^2 d\xi, \tag{66}$$

where $\xi^0 \in \gamma$. Then multiplying (66) by $(|\xi_1|^3 + 1)^{-1}$ and integrating it over $\{\xi : 0 < \xi_j < 1, j = 2, \ldots, n\}$ and over $-M < \xi_1 < M$, we obtain (64).

In order to prove the estimate.

$$\int_{\Pi_M} |\nabla W_M|^2 d\xi < K_1, \qquad K_1 = const,$$

we take $\varphi = W_M$ in (60). It is evident that

$$\left| \int_{\Pi_M} \sum_{j=1}^{n} f_j \frac{\partial W_M}{\partial \xi_j} d\xi \right| \leqslant \left(\int_{\Pi_M} \sum_{j=1}^{n} f_j^2 d\xi \right)^{1/2} \left(\int_{\Pi_M} |\nabla W_M|^2 d\xi \right)^{1/2}, \quad (67)$$

and

$$\left| \int_{\Pi_M^+} \nabla(\psi V) \nabla W_M d\xi \right| \leqslant \left(\int_{\Pi_M^+} |\nabla(\psi V)|^2 d\xi \right)^{1/2} \left(\int_{\Pi_M^+} |\nabla W_M|^2 d\xi \right)^{1/2}$$

$$(68)$$

Using (62) for W_M in Π_1 and the Friedrichs inequality in Π_1, we get

$$\left| \int_{\gamma} ([f_1] + (W + J)) W_M d\hat{\xi} \right|$$

$$\leqslant \left[\left(\int_{\gamma} [f_1]^2 d\hat{\xi} \right)^{1/2} + \left(\int_{\gamma} (W + J)^2 d\hat{\xi} \right)^{1/2} \right] \times \left(\int_{\gamma} W_M^2 d\hat{\xi} \right)^{1/2}$$

$$\leqslant K_2 \left[\left(\int_{\gamma} [f_1]^2 d\hat{\xi} \right)^{1/2} + \left(\int_{\gamma} (W + J)^2 d\hat{\xi} \right)^{1/2} \right]$$

$$\times \left(\int_{\Pi_1} |\nabla W_M|^2 d\xi \right)^{1/2}, \qquad K_2 = const. \quad (69)$$

Taking into account the inequality (65) and the condition $|f_0(\xi)| \leqslant K_3 e^{-\alpha|\xi_1|}$, where K_3, α are positive constants, we obtain

$$\left| \int_{\Pi_M} f_0 W_M d\xi \right|$$

$$\leqslant \left(\int_{\Pi_M} f_0 \left(|\xi_1|^3 + 1 \right) d\xi \right)^{1/2} \left(\int_{\Pi_M} W_M^2 \left(|\xi_1|^3 + 1 \right)^{-1} d\xi \right)^{1/2}$$

$$\leqslant K_4 \left(\int_{\Pi_M} |\nabla W_M|^2 d\xi \right)^{1/2} \quad (70)$$

From (67)–(70) and (60) with $\varphi = W_M$ we have

$$\int\limits_{\Pi_M} |\nabla W_M|^2 \mathrm{d}\xi \leqslant K_5 , \qquad (71)$$

where K_5 does not depend on M.

Now we prove that the W_M are uniformly bounded with respect to M in $L_2(\Pi_{M'})$ for any fixed M'. It follows from the imbedding theorem that

$$\int\limits_{\Pi_{M'}} |W_M|^2 \mathrm{d}\xi \leqslant K_6 \left(\int\limits_{\Pi_1} |W_M|^2 \mathrm{d}\xi + \int\limits_{\Pi_{M'}} |\nabla W_M|^2 \mathrm{d}\xi \right) \leqslant K_6,$$

where K_6 does not depend on M. Now using the diagonalization process we can find a sequence of M such that W_M converges to w in $L^2(\Pi_{M'})$ norm and ∇W_M converges weakly in $L^2(\Pi_{M'})$ for any M' as $M \to \infty$. Let us take the function φ with compact support in ξ_1 and 1-periodic with respect to $\hat{\xi}$. From (60) we obtain as a limit

$$-\int\limits_{\Pi} \nabla w \nabla \varphi \mathrm{d}\xi \;=\; -\int\limits_{\Pi} f_0 \varphi \mathrm{d}\xi + \int\limits_{\Pi} \sum_{j=1}^{n} f_j \frac{\partial \varphi}{\partial \xi_j} \mathrm{d}\xi + \int\limits_{\Pi^+} \nabla(\psi V) \nabla \varphi \mathrm{d}\xi$$

$$+ \int\limits_{\gamma} [f_1] \varphi \mathrm{d}\hat{\xi} + \int\limits_{\gamma} (W + J) \varphi \mathrm{d}\hat{\xi}. \qquad (72)$$

Let us prove now that $w(\xi) \to 0$ as $\xi_1 \to +\infty$ and $w(\xi) \to C_-$ as $\xi_1 \to -\infty$, where C_- is a constant. In the integral identity (60) for W_M we take $\varphi = W_M e^{\alpha \xi_1}$, $\alpha = const > 0$. Then we have

$$\int\limits_{\Pi_M} |\nabla W_M|^2 e^{\alpha \xi_1} \mathrm{d}\xi$$

$$\leqslant \left| \int\limits_{\Pi_M} \alpha \frac{\partial W_M}{\partial \xi_1} W_M e^{\alpha \xi_1} \mathrm{d}\xi \right| + \left| \int\limits_{\Pi_M} f_0 W_M e^{\alpha \xi_1} \mathrm{d}\xi \right|$$

$$+ \left| \int\limits_{\Pi_M} \sum_{j=1}^{n} \frac{\partial W_M}{\partial \xi_j} f_j e^{\alpha \xi_1} \mathrm{d}\xi \right| + \left| \int\limits_{\Pi_M} f_1 W_M \alpha e^{\alpha \xi_1} \mathrm{d}\xi \right|$$

$$+ \left| \int\limits_{\Pi_M^+} \nabla(\psi V) e^{\alpha \xi_1} \nabla W_M \mathrm{d}\xi \right| + \left| \int\limits_{\Pi_M^+} \alpha \nabla(\psi V) W_M e^{\alpha \xi_1} \mathrm{d}\xi \right|$$

$$+ \left| \int_\gamma \{ [f_1] + (W + J) \} W_M e^{\alpha \xi_1} d\xi \right|. \tag{73}$$

The last integral is bounded uniformly with respect to M, since

$$\int_\gamma W_M^2 d\hat{\xi} < K_7, \qquad K_7 = const, \tag{74}$$

where K_7 does not depend on M.

For $\xi_1 > 0$ we have from the Friedrichs inequality and the boundary condition $W_M = 0$ on S,

$$\int_{\Pi_{T+1} \setminus \Pi_T} (W_M)^2 d\xi \leqslant K_8 \int_{\Pi_{T+1} \setminus \Pi_T} |\nabla W_M|^2 d\xi, \qquad T \geqslant 0, \tag{75}$$

where the constant K_8 does not depend on M and, T is an integer. Therefore, multiplying (75) by $e^{\alpha T}$ and summing up with respect to T from 0 to M we get

$$\int_{\Pi_M^+} (W_M)^2 e^{\alpha \xi_1} d\xi \leqslant K_9 \int_{\Pi_M^+} |\nabla W_M|^2 e^{\alpha \xi_1} d\xi,$$

$$\Pi_M^+ = \{ \xi : \xi_1 > 0 \} \cap \Pi_M. \tag{76}$$

In order to estimate the right hand side of (73) we note that

$$\left| \int_{\Pi_M^-} \alpha \frac{\partial W_M}{\partial \xi_1} W_M e^{\alpha \xi_1} d\xi \right|$$

$$\leqslant K_{10} \left| \int_{\Pi_M^-} |\nabla W_M|^2 d\xi + \int_{\Pi_M^-} \frac{|W_M|^2}{(1 + |\xi_1|^3)} d\xi \right|$$

$$\leqslant K_{11}, \tag{77}$$

where $\Pi_M^- = \Pi_M \cap \{ \xi : \xi_1 < 0 \}$ and the constants K_{10}, K_{11} do not depend on M by virtue of (65),(71). For Π_M^+ we have

$$\left| \int_{\Pi_M^+} \alpha \frac{\partial W_M}{\partial \xi_1} W_M e^{\alpha \xi_1} d\xi \right|$$

$$\leqslant \alpha \int\limits_{\Pi_M^+} |\nabla W_M|^2 e^{\alpha \xi_1} d\xi + \alpha \int\limits_{\Pi_M^+} |W_M|^2 e^{\alpha \xi_1} d\xi$$

$$\leqslant \alpha K_{12} \int\limits_{\Pi_M^+} |\nabla W_M|^2 e^{\alpha \xi_1} d\xi, \tag{78}$$

since (76) is valid. Taking into account (65) and (76), it is easy to see that

$$\left| \int\limits_{\Pi_M} f_0 W_M e^{\alpha \xi_1} d\xi \right|$$

$$\leqslant \left| \int\limits_{\Pi_M^+} f_0 W_M e^{\alpha \xi_1} d\xi \right| + \left| \int\limits_{\Pi_M^-} f_0 W_M e^{\alpha \xi_1} d\xi \right|$$

$$\leqslant \delta \int\limits_{\Pi_M^+} |W_M|^2 e^{\alpha \xi_1} d\xi + \frac{1}{\delta} \int\limits_{\Pi_M^+} |f_0|^2 e^{\alpha \xi_1} d\xi$$

$$+ \delta_1 \int\limits_{\Pi_M^-} \frac{W_M^2 d\xi}{(1 + |\xi_1|^3)} + K_{13} \int\limits_{\Pi_M^-} |f_0|^2 e^{2\alpha \xi_1} (1 + |\xi_1|^3) d\xi$$

$$\leqslant \delta_2 \int\limits_{\Pi_M^+} |\nabla W_M|^2 e^{\alpha \xi_1} d\xi + K_{14}, \tag{79}$$

if $\alpha > 0$ is sufficiently small, δ_2 is an arbitrary positive number. The fourth integral on the right hand side of (73) can be estimated in the same way as the second integral. For the third integral we have the estimate

$$\left| \int\limits_{\Pi_M} \sum_{j=1}^n f_j \frac{\partial W_M}{\partial \xi_j} e^{\alpha \xi_1} d\xi \right| \leqslant \delta \int\limits_{\Pi_M} |\nabla W_M|^2 e^{\alpha \xi_1} d\xi + K_{15}(\delta) \int\limits_{\Pi_M} \sum_{j=1}^n f_j^2 e^{\alpha \xi_1} d\xi, \tag{80}$$

Here δ is an arbitrary small positive number. Since ψ has compact support in ξ_1, the remaining two integrals are bounded uniformly with respect to M. From (73)–(80) we have

$$\left\{ \begin{array}{l} \displaystyle\int\limits_{\Pi_M} |\nabla W_M|^2 e^{\alpha \xi_1} d\xi \leqslant K_{16}, \\[2em] \displaystyle\int\limits_{\Pi_M^+} |W_M|^2 e^{\alpha \xi_1} d\xi \leqslant K_{17}. \end{array} \right\} \tag{81}$$

It follows from (81) that

$$
\left\{
\begin{array}{l}
\displaystyle\int_{\Pi_M^+} |\nabla w|^2 e^{\alpha\xi_1}\,\mathrm{d}\xi \leqslant K_{16}, \\[6pt]
\displaystyle\int_{\Pi_M^+} |w|^2 e^{\alpha\xi_1}\,\mathrm{d}\xi \leqslant K_{17},
\end{array}
\right\}
\qquad K_{16}, K_{17} = const. \qquad (82)
$$

According to the E. De Giorgi estimate (see [7],[10])

$$
\max_{\Pi_{T+1}^+\setminus\Pi_T^+} |w(\xi)|
$$

$$
\leqslant C_8\left(\|w\|_{L_2(\Pi_{T+2}^+\setminus\Pi_{T-1}^+)} + \sum_{j=1}^n \|f_j\|_{L_2(\Pi_{T+2}^+\setminus\Pi_{T-1}^+)} \right)
$$

$$
\leqslant C_8\left(\int_{\Pi_{T+2}^+\setminus\Pi_{T-1}^+} e^{\alpha\xi_1}e^{-\alpha\xi_1}w^2\,\mathrm{d}\xi \right)^{1/2}
$$

$$
+ \sum_{j=0}^n \left(\int_{\Pi_{T+2}^+\setminus\Pi_{T-1}^+} f_j^2 e^{\alpha_0\xi_1}e^{-\alpha_0\xi_1}\,\mathrm{d}\xi \right)^{1/2}
$$

$$
\leqslant C_9 e^{-\alpha'T}, \qquad (83)
$$

where C_8, C_9 are positive constants, C_8, C_9 do not depend on T and $\alpha' > 0$ is a sufficiently small constant. Thus we have proved that

$$
N(\xi) = w(\xi) + V(\hat{\xi})\psi(\xi_1) \to 0 \text{ as } \xi_1 \to +\infty \text{ and}
$$
$$
|N(\xi)| \leqslant C_{10}e^{-\alpha'\xi_1}, C_{10}, \alpha' = const > 0, \xi_1 > 0.
$$

Consider the behavior of w when $\xi_1 \to -\infty$. The case when $f_j = 0$ for $j = 1,\ldots,n$ is considered in [5]. From the integral identity (60) with $\varphi = W_M e^{\alpha\xi_1}$, where $\alpha = const < 0$, we have

$$
-\int_{\Pi_M} |\nabla W_M|^2 e^{\alpha\xi_1}\,\mathrm{d}\xi - \int_{\Pi_M} \alpha\frac{\partial W_M}{\partial\xi_1}W_M e^{\alpha\xi_1}\,\mathrm{d}\xi
$$

$$
= -\int_{\Pi_M} f_0 W_M e^{\alpha\xi_1}\,\mathrm{d}\xi + \int_{\Pi_M} \sum_{j=1}^n f_j\frac{\partial W_M}{\partial\xi_j}e^{\alpha\xi_1}\,\mathrm{d}\xi
$$

$$+ \int\limits_{\Pi_M} f_1 W_M \alpha e^{\alpha \xi_1} \mathrm{d}\xi + \int\limits_{\Pi_M^+} \nabla(\psi V) \nabla W_M e^{\alpha \xi_1} \mathrm{d}\xi$$

$$+ \int\limits_{\Pi_M^+} \frac{\partial}{\partial \xi_1} (\psi V) W_M \alpha e^{\alpha \xi_1} \mathrm{d}\xi$$

$$+ \int\limits_{\gamma} ([f_1] + (W + J)) W_M e^{\alpha \xi_1} \mathrm{d}\hat{\xi}. \tag{84}$$

The last three integrals in the right hand side of (84) are uniformly bounded with respect to M, since ψ has compact support in ξ_1 and the W_M are uniformly bounded on γ in $L_2(\gamma)$ norm. Next, it is easy to see that

$$\left| \int\limits_{\Pi_M} f_1 W_M \alpha e^{\alpha \xi_1} \mathrm{d}\xi \right|$$

$$\leqslant |\alpha| \int\limits_{\Pi_M} \frac{W_M^2 \mathrm{d}\xi}{(1 + |\xi_1|^3)} + |\alpha| \int\limits_{\Pi_M} f_1^2 e^{2\alpha \xi_1} \left(1 + |\xi_1|^3\right) \mathrm{d}\xi \leqslant K_{18}, \tag{85}$$

$$\left| \int\limits_{\Pi_M} \sum_{j=1}^n f_j \frac{\partial W_M}{\partial \xi_j} e^{\alpha \xi_1} \mathrm{d}\xi \right| \leqslant \delta \int\limits_{\Pi_M} |\nabla W_M|^2 e^{\alpha \xi_1} \mathrm{d}\xi + K_{19} \int\limits_{\Pi_M} \sum_{j=1}^n f_j^2 e^{\alpha \xi_1} \mathrm{d}\xi, \tag{86}$$

$$\left| \int\limits_{\Pi_M} f_0 W_M e^{\alpha \xi_1} \mathrm{d}\xi \right| \leqslant \int\limits_{\Pi_M} \frac{W_M^2}{(1 + |\xi_1|^3)} \mathrm{d}\xi + \int\limits_{\Pi_M} f_0^2 e^{2\alpha \xi_1} (1 + |\xi_1|^3) \mathrm{d}\xi, \tag{87}$$

if $|\alpha|$ is sufficiently small. In order to estimate the integral

$$|\alpha| \int\limits_{\Pi_M} \frac{\partial W_M}{\partial \xi_1} W_M e^{\alpha \xi_1} \mathrm{d}\xi,$$

we use the Hardy inequality of the form

$$\int\limits_{-\infty}^{+\infty} |v - C|^2 e^{\alpha \xi_1} \mathrm{d}\xi_1$$

$$\leqslant \frac{4}{\alpha^2} \int\limits_{-\infty}^{+\infty} \left| \frac{\mathrm{d}v}{\mathrm{d}\xi_1} \right|^2 e^{\alpha \xi_1} \mathrm{d}\xi_1, \ C, \alpha = const. \tag{88}$$

This inequality follows from the classical Hardy inequality

$$\int_0^\infty t^{-2}|v|^2 dt \leqslant 4 \int_0^\infty |v'|^2 dt, \qquad v(0) = 0,$$

upon setting $t = e^{-\alpha\xi_1}$.

Taking into account (88), we get

$$|\alpha| \left| \int_{\Pi_M} \frac{\partial W_M}{\partial \xi_1} W_M e^{\alpha\xi_1} d\xi \right|$$

$$\leqslant \int_{\Pi_M} |\alpha| \left| \frac{\partial W_M}{\partial \xi_1} \right|^2 e^{\alpha\xi_1} d\xi + |\alpha| \int_{\Pi_M} |W_M|^2 e^{\alpha\xi_1} d\xi$$

$$\leqslant |\alpha| K_{20} \int_{\Pi_M} |\nabla W_M|^2 e^{\alpha\xi_1} d\xi, \tag{89}$$

since the constant C in (88) has to be taken as zero, because $W_M = 0$ for $\xi_1 = -M$ and for $\xi_1 = M$ and we extend W_M to be zero for $\xi_1 < -M$ and $\xi_1 > M$. From (84)–(89) it follows that

$$\int_{\Pi_M} |\nabla W_M|^2 e^{\alpha\xi_1} d\xi \leqslant K_{21}, \tag{90}$$

if $|\alpha|$ is sufficiently small. From (90) we have

$$\int_\Pi |\nabla w|^2 e^{\alpha\xi_1} d\xi \leqslant K_{21}. \tag{91}$$

From the Hardy inequality (88), (91) we can derive that

$$\int_\Pi |w - \overline{c}|^2 e^{\alpha\xi_1} d\xi \leqslant \overline{c}_1 \int_\Pi |\nabla w|^2 e^{\alpha\xi_1} d\xi, \qquad \overline{c}, \overline{c}_1 = const. \tag{92}$$

Indeed, from the Poincaré inequality we have

$$\int_{\gamma(\xi_1)} |w - \tilde{w}|^2 d\hat{\xi} \leqslant \overline{c}_2 \int_{\gamma(\xi_1)} |\nabla w|^2 d\hat{\xi} \quad \text{where } \tilde{w}(\xi_1) = \int_{\gamma(\xi_1)} w(\xi) d\hat{\xi}. \tag{93}$$

Multiplying (93) by $e^{\alpha\xi_1}$ and integrating over ξ_1 from $-\infty$ to ∞ we get

$$\int_\Pi |w - \tilde{w}|^2 e^{\alpha\xi_1} d\xi \leqslant \overline{c}_2 \int_\Pi |\nabla w|^2 e^{\alpha\xi_1} d\xi. \tag{94}$$

According to (88)

$$\int_\Pi |\tilde{w} - \overline{c}|^2 e^{\alpha \xi_1} \mathrm{d}\xi \leqslant \overline{c}_3 \int_\Pi |\nabla \tilde{w}|^2 e^{\alpha \xi_1} \mathrm{d}\xi \qquad (95)$$

From (94) and (95) we get (92).

Taking into account (92) and (91) we obtain

$$\int_\Pi |w - \overline{c}|^2 e^{\alpha \xi_1} \mathrm{d}\xi \leqslant \overline{c}_1 \int_\Pi |\nabla w|^2 e^{\alpha \xi_1} \mathrm{d}\xi \leqslant \overline{c}_4. \qquad (96)$$

From (96) and the E. De Giorgi theorem we obtain, as before,

$$|w(\xi) - \overline{c}| \leqslant \overline{c}_5 e^{\alpha_1 \xi_1}, \qquad \alpha_1 = const > 0, \xi_1 < 0.$$

We can take $q = \overline{c}$ in (50). Theorem 1 is proved.

Using Theorem 1 we can construct functions $N^+(\xi)$ and $N^-(\xi)$ and find constants $g_i^\pm$. We can take h_i^+ to be arbitrary constants. The constants $j_i^\pm$ also can be taken to be arbitrary according to Theorem 1. We set

$$N_0(\xi) = 1 \text{ for } \xi_1 < 0, N_0(\xi) = 0 \text{ for } \xi_1 > 0,$$
$$N_i(\xi) = 0 \text{ for } |i| \geqslant 1 \text{ and } \xi_1 < 0,$$
$$N_i(\xi) = 0 \text{ for } |i| = 1 \text{ and } \xi_1 > 0,$$
$$h_0^\pm = 0, h_{i_1}^\pm = 0, h_{i_1 i_2}^- = \delta_{i_1 i_2}, h_i^\pm = 0 \text{ for } |i| > 2,$$
$$j_i^\pm = 0 \text{ for any } i.$$

Then we get

$$N_0^\pm = 0, g_0^+ = 0, g_0^- = -1,$$

$g_{i_1}^+ = 0$ for $i_1 \geqslant 1, g_1^-$ is defined by Theorem 1, $H_{i_1}^+ = 0$ for any $i_1 = 1, \ldots, n$. Taking into account these constants we obtain from (5) that

$$\Delta v_\epsilon - f(x) = 0 \text{ in } \Omega^-. \qquad (97)$$

From (8) we obtain using the conditions $g_0^+ = 0$, $g_0^- = -1$ that

$$v_\epsilon(x) \mid_{x_1 = -0} = \epsilon \Phi_\epsilon^-(v_\epsilon) \text{ for } x \in \omega. \qquad (98)$$

We set

$$v_\epsilon(x) = 0 \text{ for } x \in \partial \Omega^- \backslash \omega, \qquad (99)$$

We also get from (5) that

$$\Delta v_\epsilon - f(x) = 0 \text{ in } \Omega^+, \tag{100}$$

$$v_\epsilon = 0 \text{ for } x \in \partial\Omega^+\backslash\omega. \tag{101}$$

Here Φ_ϵ^- is a bounded function.

From (97)–(101) we define v_k successively:

$$\Delta v_k - \delta_{k0}f = 0 \text{ in } \Omega^-, \tag{102}$$

$$v_k(x)\mid_{x_1=-0} = \Phi_k\left(v_0\mid_{x_1=\pm0}, \ldots, v_{k-1}\mid_{x_1=\pm0}\right), x \in \omega, \tag{103}$$

$$v_k(x) = 0 \text{ for } x \in \partial\Omega^-\backslash\omega, \tag{104}$$

$k = 0, 1, 2, \ldots$, and δ_{kj} is the Kronecker symbol, $\Phi_0 = 0$,

$$\Delta v_k - \delta_{k0}f = 0 \text{ in } \Omega^+, \tag{105}$$

$$v_k = 0 \text{ for } x \in \partial\Omega^+\backslash\omega. \tag{106}$$

Here Φ_k are some given functions. We note that in the problem (105),(106) there is no boundary condition on ω and the functions v_k in Ω^+ are not uniquely defined. We take any solution of (105),(106). We assume that a sufficiently smooth solution of the problem (102),(104) exists, (see [9]). Now we consider $m+1$ members of the asymptotic expansion (3),

$$\tilde{u}_\epsilon^m(x) = \sum_{k=0}^m \epsilon^k \sum_{|i|=k} \left(N_i(\xi)D^i v_\epsilon(x) + N_i^+(\xi)D^i v_\epsilon \mid_{x_1=+0} \right.$$
$$\left. + N_i^-(\xi)D^i v_\epsilon \mid_{x_1=-0} \right), \qquad \xi = \frac{x}{\epsilon}.$$

We then have

$$\Delta\tilde{u}_\epsilon^m(x) - f(x) = \epsilon^{m-2}F_\epsilon^1(x) \text{ in } \Omega_\epsilon,$$
$$\tilde{u}_\epsilon^m = 0 \text{ on } S_\epsilon, \tilde{u}_\epsilon^m = \epsilon\varphi_\epsilon(x) \text{ on } \Gamma_\epsilon,$$
$$[\tilde{u}_\epsilon^m] = 0 \text{ for } x \in \omega, \left[\frac{\partial\tilde{u}_\epsilon^m}{\partial x_1}\right] = 0 \text{ for } x \in \omega,$$

where $F_\epsilon^1, \varphi_\epsilon$ are some given functions. Therefore for

$$w_\epsilon = u_\epsilon - \tilde{u}_\epsilon^m$$

we have the following boundary-value problem:

$$\Delta w_\epsilon = \epsilon^{m-2} F_\epsilon^1 \text{ in } \Omega_\epsilon,$$
$$w_\epsilon = 0 \text{ on } S_\epsilon, w_\epsilon = \epsilon\varphi_\epsilon(x) \text{ on } \Gamma_\epsilon,$$
$$[w_\epsilon] = 0 \text{ for } x \in \omega, \left[\frac{\partial w_\epsilon}{\partial x_1}\right] = 0 \text{ for } x \in \omega.$$

In order to estimate w_ϵ we set

$$w_\epsilon = w_\epsilon^1 + w_\epsilon^2,$$

where

$$\Delta w_\epsilon^1 = \epsilon^{m-2} F_\epsilon^1(x) \text{ in } \Omega_\epsilon, \tag{107}$$
$$w_\epsilon^1 = 0 \text{ on } \Omega_\epsilon, \tag{108}$$
$$[w_\epsilon^1] = 0 \text{ for } x \in \omega \text{ and } \left[\frac{\partial w_\epsilon^1}{\partial x_1}\right] = 0 \text{ for } x \in \omega, \tag{109}$$
$$\Delta w_\epsilon^2 = 0 \text{ in } \Omega_\epsilon, \tag{110}$$
$$w_\epsilon^2 = 0 \text{ on } S_\epsilon, w_\epsilon^2 = \epsilon\varphi_\epsilon(x) \text{ on } \Gamma_\epsilon, \tag{111}$$
$$[w_\epsilon^2] = 0 \text{ for } x \in \omega, \left[\frac{\partial w_\epsilon^2}{\partial x_1}\right] = 0 \text{ for } x \in \omega. \tag{112}$$

In order to estimate w_ϵ^1 we multiply (107) by w_ϵ^1, transform the left hand side by integration by parts and apply the Friedrichs inequality

$$\int\limits_{\Omega_\epsilon} u^2 dx \leqslant C_1 \int\limits_{\Omega_\epsilon} |\nabla u|^2 dx,$$

where C_1 does not depend on ϵ. We then get the estimate

$$\|w_\epsilon^1\|_{H^1(\Omega_\epsilon)} \leqslant C_2 \epsilon^{m-2}, C_2 = const. \tag{113'}$$

For the estimation of w_ϵ^2 we use the following lemma.

Lemma 1. *Let $v(x)$ be a solution of the problem*

$$\Delta v = 0 \text{ in } \Omega_\epsilon,$$
$$v = \epsilon\varphi_\epsilon(x) \text{ on } \Gamma_\epsilon, v = 0 \text{ on } S_\epsilon,$$

where $\|\varphi_\epsilon\|_{L_2(\partial\Omega)} \leqslant C_3\sqrt{\epsilon}$ and C_3 does not depend on ϵ. Then

$$\|v\|_{L_2(\Omega_\epsilon)} \leqslant C_4 \epsilon\sqrt{\epsilon}, \qquad C_4 = const, \tag{113}$$

with C_4 not depending on ϵ.

Proof. Assume first that φ_ϵ is a smooth function and consider the problem

$$\Delta V = 0 \text{ in } \Omega, V = \epsilon\varphi_\epsilon \text{ on } \partial\Omega.$$

Let $h(x)$ be a solution of the problem

$$\Delta h = V \text{ in } \Omega,$$
$$h = 0 \text{ on } \partial\Omega.$$

According to the Green formula

$$\int_\Omega [\Delta V h - V \Delta h]\, dx = \int_{\partial\Omega} \left(\frac{\partial V}{\partial \nu}h - \frac{\partial h}{\partial \nu}V\right) ds,$$

where ν is the direction of the exterior normal to $\partial\Omega$. Since $\Delta V = 0$ in Ω, $h = 0$ on $\partial\Omega$ we have

$$-\int_\Omega V^2 dx = -\int_{\partial\Omega} \frac{\partial h}{\partial \nu}V ds.$$

and therefore

$$\int_\Omega V^2 dx \leqslant \left(\int_{\partial\Omega}\left(\frac{\partial h}{\partial \nu}\right)^2 ds\right)^{1/2}\left(\int_{\partial\Omega} V^2 ds\right)^{1/2}. \tag{114}$$

It follows from the imbedding theorem that

$$\left(\int_{\partial\Omega}\left(\frac{\partial h}{\partial \nu}\right)^2 ds\right)^{1/2}$$
$$\leqslant C_5\left(\int_\Omega\left\{h^2 + \sum_{j=1}^n\left(\frac{\partial h}{\partial \nu}\right)^2 + \sum_{i,j=1}^n\left(\frac{\partial^2 h}{\partial x_i\partial x_j}\right)^2\right\} dx\right)^{1/2}. \tag{115}$$

From [6] it is known that

$$\|h\|_{H^2(\Omega)} \leqslant C_6\|V\|_{L^2(\Omega)}, C_6 = const. \tag{116}$$

From (114)–(116) we have

$$\|V\|_{L^2(\Omega)} \leqslant C_7\epsilon\|\varphi_\epsilon\|_{L^2(\Omega)}. \tag{117}$$

Let $\varphi_\epsilon = \varphi_\epsilon^- + \varphi_\epsilon^+$, where $\varphi_\epsilon^- \leqslant 0$ and $\varphi_\epsilon^+ \geqslant 0$ on $\partial\Omega$.
Then $v = v^+ + v^-$, $V = V^+ + V^-$, where

$$\left.\begin{aligned}
\Delta v^\pm &= 0 \text{ in } \Omega_\epsilon, \\
v^\pm &= \epsilon\varphi_\epsilon^\pm \text{ on } \Gamma_\epsilon, \\
v^\pm &= 0 \text{ on } S_\epsilon,
\end{aligned}\right\} \tag{118}$$

$$\left.\begin{aligned}
\Delta V^\pm &= 0 \text{ in } \Omega, \\
V^\pm &= \epsilon\varphi_\epsilon^\pm \text{ on } \partial\Omega.
\end{aligned}\right\} \tag{119}$$

By the maximum principle,

$$0 \leqslant v^+(x) \leqslant V^+(x) \text{ in } \Omega_\epsilon,$$

and

$$V^-(x) \leqslant v^-(x) \leqslant 0 \text{ in } \Omega_\epsilon.$$

Therefore,

$$V^-(x) \leqslant v(x) \leqslant V^+(x) \text{ in } \Omega_\epsilon.$$

Taking into account (117) for φ_ϵ^- and φ_ϵ^+, we obtain

$$\begin{aligned}
\int_{\Omega_\epsilon} v^2 \mathrm{d}x &\leqslant \int_\Omega |V^+|^2 \mathrm{d}x + \int_\Omega |V^-|^2 \mathrm{d}x \\
&\leqslant 2C_8\epsilon^2 \left(\int_{\partial\Omega} |\varphi_\epsilon^+|^2 \mathrm{d}s + \int_{\partial\Omega} |\varphi_\epsilon^-|^2 \mathrm{d}s \right) \\
&\leqslant 2C_9\epsilon^3,
\end{aligned}$$

C_8, C_9 being constants and C_9 not depending on ϵ. The lemma is proved
for the case of a smooth function φ_ϵ. For $\varphi_\epsilon \in L_2(\partial\Omega)$ we can get the
inequality (113) by approximating φ_ϵ by smooth functions.

Therefore for the solution w_ϵ^2 of the problem (110)–(112) we have the
estimate

$$\|w_\epsilon^2\|_{L_2(\Omega_\epsilon)} \leqslant C_{10}\epsilon\sqrt{\epsilon}, \tag{120}$$

where the constant C_{10} does not depend on ϵ, since for $\varphi_\epsilon(x)$ in (111) we
have $\|\varphi_\epsilon\|_{L_2(\partial\Omega_\epsilon)} \leqslant C_{11}\sqrt{\epsilon}$, as it is easy to see from (3).

Now we are in a position to prove the following theorem.

Theorem 2. *For solutions $u_\epsilon(x)$ of the problem (1),(2) the estimates*

$$\|u_\epsilon(x) - v_0(x) - \epsilon v_1(x)\|_{L^2(\Omega^-)} \leqslant C\epsilon\sqrt{\epsilon}, \qquad (121)$$

$$\|u_\epsilon(x)\|_{L^2(\Omega_\epsilon^+)} \leqslant C\epsilon\sqrt{\epsilon}, \qquad (122)$$

are valid where the function $v_0(x)$ is a solution of the boundary-value problem

$$\Delta v_0 = f \ in \ \Omega^-, v_0 = 0 \ on \ \partial\Omega^-,$$

the function $v_1(x)$ is a solution of the boundary-value problem

$$\Delta v_1 = 0 \ in \ \Omega^-, v_1 = 0 \ on \ \partial\Omega^-\backslash\omega,$$

$$v_1(x) \mid_{x_1=0} = -g_1^- \frac{\partial v_0}{\partial x_1} \mid_{x_1=-0}, x \in \omega,$$

where g_1^- is a constant for which the problem (33)–(37) has a solution for $i_1 = 1$ and $j_{i_1}^- = 0$ according to Theorem 1.

Proof. The estimates (121),(122) follows from the estimates (120) for w_ϵ^2, (113′) for w_ϵ^1 and from the estimate $|N_{i_1}^\pm(\xi)| \leqslant Ce^{-\alpha|\xi_1|}$, proved in Theorem 1.

We can suppose that $v_0(x) \in H^2(\Omega^-)$, (see [9]). If $\Omega = \{x : |x_1| \leqslant l, 0 < x_j < 1, j = 2, \ldots, n\}$, $l = const > 0$, then for solutions of the equation (1) with the boundary conditions:

$u_\epsilon = \varphi_1$ for $x_1 = -l$, $u_\epsilon = \varphi_2$ for $x_1 = l$, u_ϵ is 1-periodic in $x_2, \ldots, x_n$ one can get a full asymptotic expansion by the same way as Theorem 2 is proved.

Let us consider the spectral problem, corresponding to the boundary-value problem (1),(2):

$$\Delta u_\epsilon^k + \lambda_\epsilon^k u_\epsilon^k = 0 \ in \ \Omega_\epsilon, \ 0 < \lambda_\epsilon^1 \leqslant \lambda_\epsilon^2 \leqslant \cdots. \qquad (123)$$

$$u_\epsilon^k = 0 \ on \ \partial\Omega_\epsilon, \ \int_{\Omega_\epsilon} u_\epsilon^k u_\epsilon^e dx = \delta_{ke}. \qquad (124)$$

Eigenvalues λ_ϵ^k of the problem (123),(124) form an increasing sequence and each eigenvalue is repeated as many times as its multiplicity.

Consider also the limiting eigenvalue problem:

$$\Delta u^k + \lambda^k u^k = 0 \ in \ \Omega^-, \qquad (125)$$

$$u^k = 0 \ on \ \partial\Omega^-, \qquad (126)$$

$$\int_{\Omega^-} u^k u^e dx = \delta_{ke}, \ 0 < \lambda^1 \leqslant \lambda^2 \leqslant \ldots.$$

Here, again, the λ^k form an increasing sequence and each eigenvalue is repeated as many times as its multiplicity.

We study the behavior of the eigenfunctions u_ϵ^k and eigenvalues λ_ϵ^k as $\epsilon \to 0$. Our aim is to estimate the difference between eigenvalues and eigenfunctions of the problem (123),(124) and those of the problem (125),(126) for small ϵ. For this we use the theorem for abstract operators, proved in [1] (see also [8]), on the behavior of eigenvalues and eigenvectors of a sequence of abstract operators defined in different spaces under certain restrictions imposed on this sequence.

First we formulate this theorem below.

Let H_ϵ, H_0 be separable Hilbert spaces with real valued scalar products

$$(u^\epsilon, v^\epsilon)_{H_\epsilon}, \quad (u, v)_{H_0},$$

respectively, and let

$$A_\epsilon : H_\epsilon \to H_\epsilon, \quad A_0 : H_0 \to H_0$$

be continuous linear operators, $\mathrm{Im}\, A_0 \subset V \subset H_0$, where V is a subspace of H_0 and

$$\mathrm{Im}\, A_0 = \{v : v = A_0 u, u \in H_0\}, 0 < \epsilon < 1.$$

Theorem 3. *Let the spaces H_ϵ, H_0, V and operators A_ϵ, A_0 satisfy the following conditions:*

  I. There exist linear operators $R_\epsilon : H_0 \to H_\epsilon$ and a constant $\gamma > 0$ such that

$$\left(R_\epsilon f^0, R_\epsilon f^0\right)_{H_\epsilon} \to \gamma \left(f^0, f^0\right)_{H_0} \quad as\ \epsilon \to 0$$

  for any $f^0 \in V$.

  II. The operators $A_\epsilon : H_\epsilon \to H_\epsilon$, $A_0 : H_0 \to H_0$ are positive, compact and selfadjoint; their norms $\|A_\epsilon\|_{L(H_\epsilon)}$ are bounded by a constant independent of ϵ.

  III. For any $f \in V$

$$\|A_\epsilon R_\epsilon f - R_\epsilon A_0 f\|_{H_\epsilon} \to 0 \quad as\ \epsilon \to 0.$$

  IV. The family of operators A_ϵ is uniformly compact in the following sense. From each sequence $f^\epsilon \in H_\epsilon$ such that $\|f^\epsilon\|_{H_\epsilon} < \infty$ one can extract a sub-sequence $f^{\epsilon'}$ such that for some $w^0 \in V$

$$\|A_{\epsilon'} f^{\epsilon'} - R_{\epsilon'} w^0\|_{H_{\epsilon'}} \to 0 \quad as\ \epsilon' \to 0.$$

Then for the spectral problems for operators A_ϵ, A_0

$$\left.\begin{array}{l} A_\epsilon u_\epsilon^k = \mu_\epsilon^k u_\epsilon^k, k = 1, 2, \ldots, u_\epsilon^k \in H_\epsilon, \\ \mu_\epsilon^1 \geqslant \mu_\epsilon^2 \geqslant \cdots \geqslant \mu_\epsilon^k \geqslant \ldots, \mu_\epsilon^k > 0, \\ \left(u_\epsilon^l, u_\epsilon^m\right)_{H_\epsilon} = \delta_{lm}, \end{array}\right\} \qquad (127)$$

and

$$\left.\begin{array}{l} A_0 u_0^k = \mu_0^k u_0^k, k = 1, 2, \ldots, u_0^k \in H_0, \\ \mu_0^1 \geqslant \mu_0^2 \geqslant \ldots \geqslant \mu_0^k \geqslant \ldots, \mu_0^k > 0, \\ \left(u_0^l, u_0^m\right)_{H_0} = \delta_{lm}, \end{array}\right\} \qquad (128)$$

where δ_{lm} is the Kronecker symbol, the eigenvalues μ_ϵ^k and μ_0^k, $k = 1, 2, \ldots$, form decreasing sequences and each eigenvalue is counted as many times as its multiplicity, we have

$$|\mu_\epsilon^k - \mu_0^k| \leqslant C \sup_{\substack{u \in N(\mu_0^k, A_0) \\ \|u\|_{H_0} = 1}} \|A_\epsilon R_0 u - R_\epsilon A_0 \mu\|_{H_\epsilon}, \qquad k = 1, 2, \ldots,$$

$$(129)$$

where $N(\mu_0^k, A_0) = \left\{u \in H_0, A_0 u = \mu_0^k u\right\}$ is the eigenspace of the operator A_0 corresponding to the eigenvalue μ_0^k, the constant C does not depend on ϵ, but depends on k.

Let $k \geqslant 0$, $m \geqslant 1$ be integers such that

$$\mu_0^k > \mu_0^{k+1} = \cdots = \mu_0^{k+m} > \mu_0^{k+m+1},$$

i.e. the multiplicity of the eigenvalue μ_0^{k+1} of the problem (128) is equal to m, $\mu_0^0 = \infty$. Then for any $w \in N\left(\mu_0^{k+1}, A_0\right)$, $\|w\|_{H_0} = 1$, there is a linear combination $\overline{u}_\epsilon$ of eigenvectors $u_\epsilon^{k+1}, \ldots, u_\epsilon^{k+m}$ of the problem (127) such that

$$\|\overline{u}_\epsilon - R_\epsilon w\|_{H_\epsilon} \leqslant M_k \|A_\epsilon R_\epsilon w - R_\epsilon A_0 w\|_{H_\epsilon}, \qquad (130)$$

where M_k is a constant independent of ϵ.

In order to study the behavior of the eigenfunctions and eigenvalues of the problem (127) as $\epsilon \to 0$, we apply Theorem 3.

Let us define for this case in a proper way spaces H_ϵ, H_0, V and operators A_ϵ, A_0, R_ϵ.

Denote by H_ϵ the Hilbert space consisting of all functions of $L_2(\Omega_\epsilon)$. The scalar product in H_ϵ is defined by the formula

$$(u, v)_{H_\epsilon} = \int_{\Omega_\epsilon} uv \mathrm{d}x.$$

By H_0 we denote the space of functions of $L_2(\Omega^-)$, where scalar product is given by

$$(u,v)_{H_0} = \int\limits_{\Omega^-} uv\,\mathrm{d}x$$

and $V = H^1(\Omega^-)$.

The operator $R_\epsilon : H_0 \to H_\epsilon$ we define, setting for any $f \in H_0$, $R_\epsilon f = f$ for $x \in \Omega^-$ and $R_\epsilon f = 0$ for $x \in \Omega_\epsilon^+$.

Let us define operators $A_\epsilon : H_\epsilon \to H_\epsilon$, setting $A_\epsilon f^\epsilon = -u_\epsilon$, where u_ϵ is a solution of the boundary-value problem

$$\Delta u_\epsilon = f^\epsilon \text{ in } \Omega_\epsilon, u_\epsilon = 0 \text{ on } \partial\Omega_\epsilon, \tag{131}$$

and define $A_0 : H_0 \to H_0$, setting $A_0 f^0 = -u_0$, where u_0 is a solution of the boundary-value problem

$$\Delta u_0 = f^0 \text{ in } \Omega^-, \ u_0 = 0 \text{ on } \partial\Omega^-. \tag{132}$$

Let us check that the conditions I–IV are satisfied.

I. It is evident that

$$\left(R_\epsilon f^0, R_\epsilon f^0\right)_{H_\epsilon} = \int\limits_{\Omega^-} |f^0|^2 \mathrm{d}x = \left(f^0, f^0\right)_{H_0}.$$

In our case $\gamma = 1$.

II. It is easy to check that A_ϵ and A_0 are positive, compact and selfadjoint. For the solution u_ϵ of the problem (131) we have the energy estimate, which one can get from the equality

$$(\Delta u_\epsilon, u_\epsilon)_{L_2(\Omega_\epsilon)} = (f^\epsilon, u_\epsilon)_{L_2(\Omega_\epsilon)}, \tag{133}$$

transforming the left hand side by integration by parts and applying the Friedrichs inequality in the right hand side of (133). We get

$$\left(A_\epsilon f^\epsilon, A_\epsilon f^\epsilon\right)_{H_\epsilon} = \|u_\epsilon\|_{L_2(\Omega_\epsilon)}^2 \leqslant C_1 \|\nabla u_\epsilon\|_{L^2(\Omega_\epsilon)}^2 \leqslant C_2 \|f^\epsilon\|_{H_\epsilon}^2.$$

III. We have to prove that

$$\|u_\epsilon - \tilde{u}_0\|_{H_\epsilon} \to 0 \text{ as } \epsilon \to 0, \tag{134}$$

where u_ϵ is a solution of the problem (131) with $f^\epsilon = f$ in Ω^- and $f^\epsilon = 0$ in Ω_ϵ^+, $\tilde{u}_0$ is a solution of problem (132) in Ω^- with $f^0 = f$ and $\tilde{u}_0 = 0$ in Ω_ϵ^+. The relation (134) is valid, since

$$\|u_\epsilon - \tilde{u}_0\|_{L^2(\Omega^-)} \to 0 \text{ and } \|u_\epsilon\|_{L^2(\Omega_\epsilon^+)} \to 0$$

according to the estimates (121) and (122).

IV. By the energy estimate and the Friedrichs inequality we have $\|A_\epsilon f^\epsilon\|_{H^1(\Omega_\epsilon)} \leqslant C_1 \|f^\epsilon\|_{H_\epsilon}$, C_1 does not depend on ϵ, and therefore according to the Rellich–Sobolev compactness theorem there is a sequence $f^{\epsilon'}$ such that

$$\|u^{\epsilon'} - w^0\|_{L_2(\Omega^-)} + \|u^{\epsilon'}\|_{L_2(\Omega_{\epsilon'}^+)} \to 0 \text{ as } \epsilon' \to 0,$$

where $u^{\epsilon'}$ is a solution of the problem

$$\Delta u^{\epsilon'} = f^{\epsilon'} \text{ in } \Omega_{\epsilon'}, \quad u^{\epsilon'} = 0 \text{ on } \partial\Omega_{\epsilon'}.$$

Thus we have (129) and (130). It is easy to see that $\mu_\epsilon^k = (\lambda_\epsilon^k)^{-1}$, $\mu_0^k = (\lambda^k)^{-1}$. We have proved the following theorem for the eigenvalue problem (123),(124).

Theorem 4. *Let λ_ϵ^k be an eigenvalue of the problem (123),(124) and λ^k be an eigenvalue of the problem (125),(126). Then*

$$\left| \frac{1}{\lambda_\epsilon^k} - \frac{1}{\lambda^k} \right| \leqslant C(k)\epsilon,$$

where $C(k)$ does not depend on ϵ.

Let $k \geqslant 0$, $m \geqslant 1$ be integers such that

$$\frac{1}{\lambda^k} > \frac{1}{\lambda^{k+1}} = \cdots = \frac{1}{\lambda^{k+m}} > \frac{1}{\lambda^{k+m+1}},$$

i.e. the multiplicity of the eigenvalue λ^{k+1} is equal to m, $\mu_0^0 = \infty$. Then for any w which is an eigenfunction corresponding to the eigenvalue λ^{k+1} with $\|w\|_{H_0} = 1$, there is a linear combination $\overline{u}_\epsilon$ of eigenfunctions $u_\epsilon^{k+1}, \ldots, u_\epsilon^{k+m}$ of the problem (123),(124) such that

$$\|\overline{u}_\epsilon - R_\epsilon w\|_{H_\epsilon} \leqslant M_k \epsilon,$$

where M_k is a constant independent of ϵ.

The proof of Theorem 4 is an application of Theorem 3 to the problem (127),(128).

The results of this section were obtained jointly with A.S. Shamaev.

160   Chapter 4.   On some homogenization problems

References

[1] O.A. Oleinik, A.S. Shamaev, G.A. Yosifian, Mathematical problems in elasticity and homogenization, North-Holland, Amsterdam, 1992.

[2] V.V. Jikov, S.M. Kozlov, O.A. Oleinik , Homogenization of differential operators and integral functionals, Springer Verlag, Heidelberg, 1994.

[3] W. Jäger, A. Mikelic, Homogenization of the Laplace equation in a partially perforated domain, Preprint, University of Heidelberg, 1993.

[4] K. Yosida, Functional Analysis, Springer Verlag, 1965.

[5] O.A. Oleinik, G.A. Yosifian, On the behavior at infinity of solutions of second order elliptic equations in domains with a noncompact boundary, Math. Sbornik 112, 1980, No 4, 588–610.

[6] L. Bers, F. John, M. Schechter, Partial differential equations, Interscience Publishers, New York, 1964.

[7] E. De Giorgi, Sulla differenziabilità e l'analiticità delle estremali degli integrali, Mem. Acc. Sci. Torino 1957, 1–19.

[8] O.A. Oleinik, A.S. Shamaev, G.A. Yosifian, On a limit behavior of the spectrum of a sequence of operators which are defined in different Hilbert spaces, Russian Math. Surveys 44, 1989, No 3.

[9] V.A. Kondratiev, On the smoothness of solutions of the Dirichlet problem for an elliptic second order equation in a piecewise smooth domain, Diff. equations 6, 1970, 10, 1831–1843.

[10] D. Gilbarg, N.S. Trudinger, Elliptic partial differential equations of second order, 2nd edition, Springer Verlag, Berlin, 1983.

[11] A. Bensoussan, J.-L. Lions, G. Papanicolau, Asymptotic analysis for periodic structures, North Holland, Amsterdam, 1978.

[12] O.A. Oleinik, A.S. Shamaev, On homogenization of solutions of a boundary-value problem for the Laplace operator in a partially perforated domain with the Dirichlet condition on the boundary of cavities, Doklady RAS 337, 1994, No 2, 168–171.

[13] O.A. Oleinik, T.A. Shaposhnikova, On an approach to construct approximations for the homogenization problem in a partially perforated domain, Diff. equations 30, 1994, No 11, 1994-1999.

[14] O.A. Oleinik, T.A. Shaposhnikova, On homogenization problem in a partially perforated domain with the mixed boundary condition on the boundary of cavities, containing a small parameter, Diff. equations, No. 7, 1995.

4.2 On a homogenization problem for the Laplace operator in a partially perforated domain with the Neumann condition on holes

In this section we use the same notation as in section 1 and a similar approach. We study the boundary-value problem in the partially perforated domain Ω_ϵ,

$$\Delta u_\epsilon = f \text{ in } \Omega_\epsilon, f \in C^\infty(\overline{\Omega}^+), f \in C^\infty(\overline{\Omega}^-), \tag{1}$$

$$\frac{\partial u_\epsilon}{\partial \nu} = 0 \text{ on } S_\epsilon, u_\epsilon = 0 \text{ on } \Gamma_\epsilon, u \in H^1(\Omega_\epsilon), \tag{2}$$

where ν is the exterior normal direction on S_ϵ. We look for an asymptotic expansion for u_ϵ of the form

$$\tilde{u}_\epsilon(x) \approx \sum_{k=0}^\infty \epsilon^k \sum_{|i|=k} \left(N_i(\xi) D^i v_\epsilon(x) + N_i^+(\xi) D^i v_\epsilon \mid_{x_1=+0} \right.$$
$$\left. + N_i^-(\xi) D^i v_\epsilon \mid_{x_1=-0} \right), \tag{3}$$

where $\xi = \frac{x}{\epsilon}$, $i = (i_1, \ldots, i_k)$, $D^i = \frac{\partial^{|i|}}{\partial x_{i_1}, \ldots, x_{i_k}}$, $i_j = 1, \ldots, n$, for $|i| > 0$ and $D^i v \equiv v$ for $|i| = 0$,

$$v_\epsilon(x) = v_0(x) + \epsilon v_1(x) + \epsilon^2 v_2(x) + \cdots, \tag{4}$$

$N_i(\xi)$ is a smooth 1-periodic function in $\{\xi : \xi_1 \leqslant 0\}$ and also a smooth 1-periodic function in $\{\xi : \xi_1 \geqslant 0\} \backslash g$, N_i^+, N_i^- are functions of the boundary layer type, which tend to zero as $|\xi_1| \to \infty$. They are defined in $\mathbf{R}^n \backslash g$ and are 1-periodic with respect to $\hat{\xi} = (\xi_2, \ldots, \xi_n)$, $N_i(\xi)$, $N_i^+(\xi)$, $N_i^-(\xi)$ can be discontinuous at $\xi_1 = 0$.

We substitute the expansion (3) into the equation (1), into the boundary conditions (2) on S_ϵ, and we require that the following transmission conditions are satisfied on ω:

$$\frac{\partial \tilde{u}_\epsilon}{\partial x_1}\bigg|_{x_1=+0} = \frac{\partial \tilde{u}_\epsilon}{\partial x_1}\bigg|_{x_1=-0}, \tag{5}$$

$$\tilde{u}_\epsilon|_{x_1=+0} = \tilde{u}_\epsilon|_{x_1=-0}.$$

It is easy to see that we get

$$\Delta \tilde{u}_\epsilon \approx \sum_{k=0}^\infty \epsilon^{k-2} \sum_{|i|=k} \left(H_i(\xi) D^i v_\epsilon(x) + H_i^+(\xi) D^i v_\epsilon(x) \mid_{x_1=+0} \right.$$

$$+H_i^-(\xi)D^i v_\epsilon(x)\,|_{x_1=-0}\big)\quad \text{in }\Omega_\epsilon, \tag{6}$$

$$\frac{\partial \tilde u_\epsilon}{\partial \nu}\;\approx\;\sum_{k=0}^{\infty}\epsilon^{k-1}\sum_{|i|=k}\big(B_i(\xi)D^i v_\epsilon(x)+B_i^+(\xi)D^i v_\epsilon(x)\,|_{x_1=+0}$$

$$+B_i^-(\xi)D^i v_\epsilon(x)\,|_{x_1=-0}\big)\quad \text{on }S_\epsilon, \tag{7}$$

$$\left[\frac{\partial \tilde u_\epsilon}{\partial \nu}\right]\;\approx\;\sum_{k=0}^{\infty}\epsilon^{k-1}\sum_{|i|=k}\Big(J_i^+(\hat\xi)D^i v_\epsilon(x)\,|_{x_1=+0}+J_i^-(\hat\xi)D^i v_\epsilon(x)\,|_{x_1=-0}\Big)$$

$$\text{on }\omega, \tag{8}$$

$$[\tilde u_\epsilon]\;\approx\;\sum_{k=0}^{\infty}\epsilon^{k}\sum_{|i|=k}\Big(G_i^+(\hat\xi)D^i v_\epsilon(x)\,|_{x_1=+0}+G_i^-(\hat\xi)D^i v_\epsilon(x)\,|_{x_1=-0}\Big)$$

$$\text{on }\omega. \tag{9}$$

Here and in what follows we use the notation

$$[\zeta]=\zeta\,|_{x_1=+0}-\zeta\,|_{x_1=-0}\,.$$

For the construction of the asymptotic expansion for u_ϵ we choose functions $N_i(\xi)$, $N_i^+(\xi)$, $N_i^-(\xi)$ in the following way:

$$B_i(\xi)\;=\;0,\,B_i^+(\xi)=0,\,B_i^-(\xi)=0\;\text{for any }i, \tag{10}$$

$$H_0(\xi)\;=\;0,\,H_{i_1}(\xi)=0,\,i_1=1,\ldots,n, \tag{11}$$

$$\left.\begin{array}{rcl}H_i(\xi)&=&h_i^+\;\text{for any }i\text{ with }|i|\geqslant 2\text{ and for }\xi_1>0,\\[2pt]H_i(\xi)&=&h_i^-\;\text{for any }i\text{ with }|i|\geqslant 2\text{ and for }\xi_1<0,\end{array}\right\} \tag{12}$$

$$H_i^+(\xi)\;=\;0,\,H_i^-(\xi)=0\;\text{for any }i\text{ and for }\xi\in\mathbf{R}^n, \tag{13}$$

$$J_i^+(\hat\xi)\;=\;j_i^+\;\text{for any }i\,, \tag{14}$$

$$J_i^-(\hat\xi)\;=\;j_i^-\;\text{for any }i\,, \tag{15}$$

$$G_i^+(\hat\xi)\;=\;g_i^+\;\text{for any }i\,, \tag{16}$$

$$G_i^-(\hat\xi)\;=\;g_i^-\;\text{for any }i\,, \tag{17}$$

where h_i^+, h_i^-, j_i^+, j_i^-, g_i^+, g_i^- are some constants, which we choose later. Calculations show that equations (10)–(17) lead to the following boundary-value problems for the functions $N_i(\xi)$, $N_i^+(\xi)$, $N_i^-(\xi)$.

For $\xi_1 > 0$ and for $N_i(\xi)$ we have

$$
\left\{
\begin{array}{ll}
\Delta N_0 = 0 \text{ in } Y, & (18) \\
\left.\dfrac{\partial N_0}{\partial \nu}\right|_{S_0} = 0, N_0(\xi) \text{ is 1-periodic in } \xi, & (19)
\end{array}
\right.
$$

$$
\left\{
\begin{array}{ll}
\Delta N_{i_1} + 2D^{i_1}N_0 = 0 \text{ in } Y, i_1 = 1, \ldots, n, & (20) \\
\left.\dfrac{\partial N_{i_1}}{\partial \nu}\right|_{S_0} = -N_0\nu_{i_1}\big|_{S_0}, & (21) \\
N_{i_1}(\xi) \text{ is 1-periodic in } \xi, & (22)
\end{array}
\right.
$$

$$
\left\{
\begin{array}{ll}
\Delta N_{i_1 i_2} + 2D^{i_2}N_{i_1} + N_0\delta_{i_1 i_2} = h_{i_1 i_2}^+ \text{ in } Y, & (23) \\
\left.\dfrac{\partial N_{i_1 i_2}}{\partial \nu}\right|_{S_0} = -N_{i_1}\nu_{i_2}\big|_{S_0}, & (24) \\
N_{i_1 i_2}(\xi) \text{ is 1-periodic in } \xi, i_1, i_2 = 1, \ldots, n, & (25)
\end{array}
\right.
$$

$$
\left\{
\begin{array}{ll}
\Delta N_i + 2D^{i_k}N_{i_1 \ldots i_{k-1}} + \delta_{i_{k-1} i_k}N_{i_1 \ldots i_{k-2}} = h_i^+ \text{ in } Y, & (26) \\
\left.\dfrac{\partial N_i}{\partial \nu}\right|_{S_0} = -N_{i_1 \ldots i_{k-1}}\nu_{i_k}\big|_{S_0}, & (27) \\
N_i(\xi) \text{ is 1-periodic in } \xi \text{ for any } i \text{ and } i_j = 1, \ldots, n, & (28)
\end{array}
\right.
$$

where δ_{ks} is the Kronecker symbol: $\delta_{ks} = 1$ for $k = s$ and $\delta_{ks} = 0$ for $k \neq s$.

For $\xi_1 < 0$ for functions N_i we have equations of the form (18), (20), (23),(26) in $Q = \{\xi : 0 < \xi_j < 1, j = 1, \ldots, n\}$, where the constants h_i^+ are replaced by h_i^-.

The functions $N_i^+(\xi)$ and $N_i^-(\xi)$ which have the boundary layer type are defined in $\mathbf{R}^n \backslash g$ as follows:

$$
\left\{
\begin{array}{ll}
\Delta N_0^{\pm} = 0 \text{ in } \mathbf{R}^n \backslash g, & (29) \\
\left.\dfrac{\partial N_0^{\pm}}{\partial \nu}\right|_S = 0, & (30) \\
\left[N_0^{\pm}\right] = \mp N_0\big|_{\xi_1 = \pm 0} + g_0^{\pm}, & (31) \\
\left[\dfrac{\partial N_0^{\pm}}{\partial \xi_1}\right] = \mp \left(\dfrac{\partial N_0}{\partial \xi_1}\right)\Big|_{\xi_1 = \pm 0} + j_0^{\pm}, & (32) \\
|N_0^{\pm}(\xi)| \to 0 \text{ as } |\xi_1| \to \infty, N_0^{\pm}(\xi) \text{ is 1-periodic in } \hat{\xi}, & \\
\hat{\xi} = (\xi_2, \ldots, \xi_n), & (33)
\end{array}
\right.
$$

$$\left\{ \begin{array}{ll} \Delta N_{i_1}^{\pm} + 2(1 - \delta_{i_1 1})D^{i_1}N_0^{\pm} = 0 \text{ in } \mathbf{R}^n \backslash g, & (34) \\[2mm] \left. \frac{\partial N_{i_1}^{\pm}}{\partial \nu} \right|_S = -(1 - \delta_{i_1 1})N_0^{\pm}\nu_{i_1}\big|_S\,, & (35) \\[2mm] \left[N_{i_1}^{\pm} \right] = \mp N_{i_1}|_{\xi_1 = \pm 0} + g_{i_1}^{\pm}, & (36) \\[2mm] \left[\frac{\partial N_{i_1}^{\pm}}{\partial \xi_1} \right] = \mp \left(\frac{\partial N_{i_1}}{\partial \xi_1} + \delta_{i_1 1}N_0 \right)\Big|_{\xi_1 = \pm 0} + j_{i_1}^{\pm}, & (37) \\[2mm] \left| N_{i_1}^{\pm}(\xi) \right| \to 0 \text{ as } |\xi_1| \to \infty, & (38) \\[1mm] N_{i_1}^{\pm}(\xi) \text{ is 1-periodic in } \hat{\xi}, i_1 = 1,\dots,n, & (39) \end{array} \right.$$

$$\left\{ \begin{array}{ll} \Delta N_{i_1 i_2}^{\pm} + (1 - \delta_{i_1 1})(2D^{i_1}N_{i_2}^{\pm} + N_0^{\pm}\delta_{i_1 i_2}) = 0 \text{ in } \mathbf{R}^n \backslash g, & (40) \\[2mm] \left. \frac{\partial N_{i_1 i_2}^{\pm}}{\partial \nu} \right|_S = -(1 - \delta_{i_1 1})N_{i_2}^{\pm}\nu_{i_1}\big|_S\,, & (41) \\[2mm] \left[N_{i_1 i_2}^{\pm} \right] = \mp N_{i_1 i_2}|_{\xi_1 = \pm 0} + g_{i_1 i_2}^{\pm}, & (42) \\[2mm] \left[\frac{\partial N_{i_1 i_2}^{\pm}}{\partial \xi_1} \right] = \mp \left(\frac{\partial N_{i_1 i_2}}{\partial \xi_1} + \delta_{i_1 1}N_{i_2} \right)\Big|_{\xi_1 = \pm 0} + j_{i_1 i_2}^{\pm}, & (43) \\[2mm] \left| N_{i_1 i_2}^{\pm}(\xi) \right| \to 0 \text{ as } |\xi_1| \to \infty, i_1, i_2 = 1,\dots,n, & (44) \\[1mm] N_{i_1 i_2}^{\pm}(\xi) \text{ is 1-periodic in } \hat{\xi}, i_1, i_2 = 1,\dots,n. & (45) \end{array} \right.$$

For any i we have

$$\left\{ \begin{array}{ll} \Delta N_i^{\pm} + (1 - \delta_{i_1 1})(2D^{i_1}N_{i_2 \cdots i_k}^{\pm} + N_{i_3 \cdots i_k}^{\pm}\delta_{i_1 i_2}) = 0 \text{ in } \mathbf{R}^n \backslash g, & (46) \\[2mm] \left. \frac{\partial N_i^{\pm}}{\partial \nu} \right|_S = -(1 - \delta_{i_1 1})N_{i_2 \cdots i_k}^{\pm}\nu_{i_1}\big|_S\,, & (47) \\[2mm] \left[N_i^{\pm} \right] = \mp N_i|_{\xi_1 = \pm 0} + g_i^{\pm}, & (48) \\[2mm] \left[\frac{\partial N_i^{\pm}}{\partial \xi_1} \right] = \mp \left(\frac{\partial N_i}{\partial \xi_1} + \delta_{i_1 1}N_{i_2 \cdots i_k} \right)\Big|_{\xi_1 = \pm 0} + j_i^{\pm}, & (49) \\[2mm] \left| N_i^{\pm}(\xi) \right| \to 0 \text{ as } |\xi_1| \to \infty, & (50) \\[1mm] N_i^{\pm}(\xi) \text{ is 1-periodic in } \hat{\xi}. & (51) \end{array} \right.$$

The boundary-value problems for functions $N_i(\xi)$, $N_i^{\pm}(\xi)$ have solutions only for some constants $h_i^{\pm}$, $g_i^{\pm}$, $j_i^{\pm}$.

The conditions for the existence of solutions $N_i(\xi)$ are well-known (see, for example [4]). They have the form

$$h_i^+ = \frac{1}{|Y|} \int_Y \left(D^{i_k}N_{i_1 \cdots i_{k-1}}(\xi) + \delta_{i_{k-1} i_k}N_{i_1 \cdots i_{k-2}}(\xi) \right) d\xi. \qquad (52)$$

It is easy to see that for $\xi_1 < 0$ we can take $h_i^- = 0$ for $|i| > 2$, $h_{i_1 i_2}^- = \delta_{i_1 i_2}$, $h_i^- = 0$ for $|i| < 2$, since N_i can be chosen as constants: $N_0 = 1$, $N_i = 0$ for $|i| \geqslant 1$.

In order to solve the problems for $N_i^\pm$ we use the following theorem:

Theorem 1. *There exist constants q, J such that the problem*

$$\Delta N = -f_0 - \sum_{j=1}^n \frac{\partial f_j}{\partial \xi_j} \text{ in } \mathbf{R}^n \backslash g, \tag{53}$$

$$\left.\frac{\partial N}{\partial \nu}\right|_S = -\sum_{j=1}^n \nu_j f_j \bigg|_S, \tag{54}$$

$$[N] = V(\hat{\xi}) + q \text{ on } \gamma, \tag{55}$$

$$\left[\frac{\partial N}{\partial \xi_1}\right] = W(\hat{\xi}) + J \text{ on } \gamma, \tag{56}$$

where $|f_j(\xi)| \leqslant C_1 e^{-\alpha|\xi_1|}$, $j = 0, 1, \ldots, n$, $\alpha = \text{const} > 0$, $f_0(\xi)$, $f_j(\xi)$ are 1-periodic in ξ, smooth in $\{\mathbf{R}^n \backslash g\} \cap \{\xi : \xi_1 \geqslant 0\}$ and $\mathbf{R}^n \cap \{\xi : \xi_1 \leqslant 0\}$,

$$|N(\xi)| \to 0 \text{ as } |\xi_1| \to \infty, \qquad N(\xi) \text{ is 1-periodic in } \hat{\xi},$$

has a solution $N(\xi)$ and

$$J = -\int_\gamma W(\hat{\xi})\mathrm{d}\hat{\xi} + \int_\Pi f_0(\xi)\mathrm{d}\xi - \int_\gamma [f_1]\mathrm{d}\hat{\xi}, \tag{57}$$

where $\Pi = \{\mathbf{R}^n \backslash g\} \cap \{\xi : 0 < \xi_j < 1, j = 2, \ldots, n\}$, $\gamma = \Pi \cap \{\xi : \xi_1 = 0\}$. Moreover, there exist constants β and C_2 such that

$$|N(\xi)| \leqslant C_2 e^{-\beta|\xi_1|}.$$

Proof. For a positive constant M, let

$$P_M = (\mathbf{R}^n \backslash g) \cap \{\xi : |\xi_1| < M\}, \text{ and } \Pi_M = \Pi \cap \{\xi : |\xi_1| < M\}.$$

Consider the following boundary-value problem:

$$\Delta U_M = -f_0 - \sum_{j=1}^n \frac{\partial f_j}{\partial \xi_j} \text{ in } P_M, \tag{58}$$

$$\left.\frac{\partial U_M}{\partial \nu}\right|_{S_M} = -\sum_{j=1}^n \nu_j f_j \bigg|_{S_M}, S_M = S \cap P_M, \tag{59}$$

$$[U_M] = V(\hat{\xi}) \text{ on } \gamma, \tag{60}$$

$$\left[\frac{\partial U_M}{\partial \xi_1}\right] = W(\hat{\xi}) + J_M \text{ on } \gamma, \tag{61}$$

$$\frac{\partial U_M}{\partial \xi_1} = 0 \text{ for } \xi_1 = -M, \tag{62}$$

$$\frac{\partial U_M}{\partial \xi_1} = 0 \text{ for } \xi_1 = M, \tag{63}$$

and $U_M(\xi)$ is 1-periodic in $\hat{\xi}$.

We set

$$W_M(\xi) = U_M - V(\hat{\xi})\psi(\xi_1),$$

where $\psi(\xi_1)$ is such that $\psi(\xi_1) = 1$ for $0 \leqslant \xi_1 < \delta$, $\psi(\xi_1) = 0$ for $\xi_1 < 0$, $\psi(\xi_1) = 0$ for $\xi_1 \geqslant 2\delta$, $0 \leqslant \psi \leqslant 1$, $\psi \in C^\infty(\mathbf{R}_+^1)$, and δ is a sufficiently small constant, $\mathbf{R}_+^1 = \{\xi_1 : \xi_1 \geqslant 0\}$, $\psi = 0$ on S.

The function $U_M \in H^1(\Pi_M)$ is called a weak solution of the problem (58)–(63), if the function $W_M(\xi)$ for any $\varphi \in H^1(\Pi_M)$ which is 1-periodic in $\hat{\xi}$ satisfies the integral identity

$$-\int_{\Pi_M} \nabla W_M \nabla \varphi \mathrm{d}\xi$$

$$= -\int_{\Pi_M} f_0 \varphi \mathrm{d}\xi + \int_{\Pi_M} \sum_{j=1}^n f_j \frac{\partial \varphi}{\partial \xi_j} \mathrm{d}\xi$$

$$+ \int_{\Pi_M^+} \nabla(\psi V)\nabla \varphi \mathrm{d}\xi + \int_\gamma [f_1]\varphi \mathrm{d}\hat{\xi} + \int_\gamma (W + J_M)\varphi \mathrm{d}\hat{\xi}, \tag{64}$$

where $\Pi_M^+ = \Pi_M \cap \{\xi : \xi_1 > 0\}$, $\nabla u = \operatorname{grad} u \equiv \left(\dfrac{\partial u}{\partial \xi_1}, \cdots, \dfrac{\partial u}{\partial \xi_n}\right)$.

Let us consider the space

$$H_M = \{u : u \in H^1(\Pi_M), \int_{\Pi_M} u \mathrm{d}\xi = 0, u \text{ is 1-periodic in } \hat{\xi}\}.$$

Due to the Poincaré inequality we can introduce in H_M the scalar product

$$\int_{\Pi_M} \nabla u \nabla v \mathrm{d}\xi$$

and the norm

$$\left(\int\limits_{\Pi_M} |\nabla u|^2 \mathrm{d}\xi \right)^{1/2}.$$

Therefore for the proof of the existence of a weak solution of the problem (58)–(63) it is sufficient in view of the Lax–Milgram theorem to prove that the right hand side of (64) is a linear bounded functional for φ in H_M. We have

$$\left| -\int\limits_{\Pi_M} f_0 \varphi \mathrm{d}\xi + \int\limits_{\Pi_M} \sum_{j=1}^n f_j \frac{\partial \varphi}{\partial \xi_j} \mathrm{d}\xi + \int\limits_{\Pi_M^+} \nabla(\psi V)\nabla\varphi \mathrm{d}\xi + \int\limits_\gamma [f_1]\,\varphi \mathrm{d}\hat\xi \right.$$
$$\left. + \int\limits_\gamma (W + J_M)\varphi \mathrm{d}\hat\xi \right| \leqslant C_3 \left(\int\limits_{\Pi_M} |\nabla\varphi|^2 \mathrm{d}\xi \right)^{1/2},$$

since from the Poincaré inequality,

$$\int\limits_{\Pi_M} |\varphi|^2 \mathrm{d}\xi \leqslant C_4 \int\limits_{\Pi_M} |\nabla\varphi|^2 \mathrm{d}\xi$$

and according to the imbedding theorem

$$\int\limits_\gamma |\varphi|^2 \mathrm{d}\hat\xi \leqslant C_5 \int\limits_{\Pi_M} |\nabla\varphi|^2 \mathrm{d}\xi,$$

where C_3, C_4, C_5 do not depend on φ.

Therefore, the integral identity (64) is satisfied for any $\varphi \in H_M$. In order to have (64) satisfied for any $\varphi \in H^1(\Pi_M)$ that is 1-periodic in $\hat\xi$, we need to choose a constant J_M in such a way that

$$\int\limits_\gamma (W + J_M)\mathrm{d}\hat\xi + \int\limits_\gamma [f_1]\mathrm{d}\hat\xi - \int\limits_{\Pi_M} f_0 \mathrm{d}\xi = 0. \tag{65}$$

We notice that (65) tends to (57) as $M \to \infty$.

It is easy to see that the solution of the problem (58)–(63) has the form $W_M + C$, where C is an arbitrary constant.

Now we shall prove that

$$\int\limits_{\Pi_M} |\nabla W_M|^2 \mathrm{d}\xi < K, \tag{66}$$

where the constant K does not depend on M.

We denote by $\widetilde{W}_M$ an extension of the function W_M on $\mathbf{R}^n$ such that

$$\int_{\widetilde{\Pi}_M} |\nabla \widetilde{W}_M|^2 d\xi \leqslant C_6 \int_{\Pi_M} |\nabla W_M|^2 d\xi, \tag{67}$$

where C_6 does not depend on M; $\widetilde{\Pi}_M = \{\xi : \xi \in \mathbf{R}^n, 0 < \xi_j < 1, j = 2, \ldots, n, |\xi_1| < M\}$. Such an extension is possible due to the periodic structure of $\{\xi : \xi_1 \geqslant 0\} \backslash g$ (see [1]).

For any M let us fix the solution $W_M + C_M$ of the problem (58)–(63) in such a way that

$$\int_{\Pi_1} (\widetilde{W}_M(\xi) + C_M) d\xi = 0.$$

Using the Poincaré inequality, we have

$$\int_{\widetilde{\Pi}_1} \left|\widetilde{W}_M(\xi) + C_M\right|^2 d\xi \leqslant K_1 \int_{\widetilde{\Pi}_1} \left|\nabla \widetilde{W}_M\right|^2 d\xi \tag{68}$$

where the constant K_1 does not depend on M.

From (68) and the imbedding theorem it follows that

$$\int_{\gamma} (\widetilde{W}_M + C_M)^2 d\hat{\xi} \leqslant C_7 \int_{\widetilde{\Pi}_1} \left|\nabla \widetilde{W}_M\right|^2 d\xi, \tag{69}$$

where C_7 does not depend on M. It is easy to see that

$$\int_{\widetilde{\Pi}_M} (\widetilde{W}_M + C_M)^2 \left(|\xi_1|^3 + 1\right)^{-1} d\xi$$

$$\leqslant 4 \int_{\gamma} (\widetilde{W}_M + C_M)^2 d\hat{\xi} \int_0^M \left(|\xi_1|^3 + 1\right)^{-1} d\xi_1$$

$$+ 4 \int_{\widetilde{\Pi}_M} \left|\frac{\partial \widetilde{W}_M}{\partial \xi_1}\right|^2 d\xi \int_0^M |\xi_1| \left(|\xi_1|^3 + 1\right)^{-1} d\xi_1$$

and therefore from (69) we have

$$\int_{\widetilde{\Pi}_M} \left(\widetilde{W}_M + C_M\right)^2 \left(|\xi_1|^3 + 1\right)^{-1} d\xi \leqslant C_8 \int_{\widetilde{\Pi}_M} \left|\nabla \widetilde{W}_M\right|^2 d\xi \tag{70}$$

with the constant C_8 independent of M. In order to estimate

$$\int_{\Pi_M} |\nabla W_M|^2 d\xi$$

we take $\varphi = W_M + C_M$ in (64).

It is evident that

$$\left| \int_{\Pi_M} \sum_{j=1}^n f_j \frac{\partial W_M}{\partial \xi_j} d\xi \right|$$

$$\leqslant \left(\int_{\Pi_M} \sum_{j=1}^n f_j^2 d\xi \right)^{1/2} \left(\int_{\Pi_M} |\nabla W_M|^2 d\xi \right)^{1/2}, \tag{71}$$

$$\left| \int_{\Pi_M} \nabla(\psi V) \nabla W_M d\xi \right|$$

$$\leqslant \left(\int_{\Pi_M} |\nabla(\psi V)|^2 d\xi \right)^{1/2} \left(\int_{\Pi_M} |\nabla W_M|^2 d\xi \right)^{1/2}. \tag{72}$$

Using (69) we get

$$\left| \int_\gamma ([f_1] + (W + J_M))(W_M + C_M) d\hat\xi \right|$$

$$\leqslant \left[\left(\int_\gamma [f_1]^2 d\hat\xi \right)^{1/2} + \left(\int_\gamma (W + J_M)^2 d\hat\xi \right)^{1/2} \right]$$

$$\times \left(\int_\gamma |W_M + C_M|^2 d\hat\xi \right)^{1/2}$$

$$\leqslant K_2 \left[\left(\int_\gamma [f_1]^2 d\hat\xi \right)^{1/2} + \left(\int_\gamma (W + J_M)^2 d\hat\xi \right)^{1/2} \right]$$

$$\times \left(\int_{\Pi_1} |\nabla \widetilde{W}_M|^2 d\xi \right)^{1/2}. \tag{73}$$

Taking into account the inequality (70) and the condition $|f_0(\xi)| \leqslant C_1 e^{-\alpha|\xi_1|}$, C_1, α positive constants, we obtain

$$
\left| \int\limits_{\Pi_M} f_0(W_M + C_M)\mathrm{d}\xi \right|
$$

$$
\leqslant \left(\int\limits_{\Pi_M} f_0^2(\xi)(|\xi_1|^3 + 1)\mathrm{d}\xi \right)^{1/2}
$$

$$
\times \left(\int\limits_{\Pi_M} (W_M + C_M)^2(|\xi_1|^3 + 1)^{-1}\mathrm{d}\xi \right)^{1/2}
$$

$$
\leqslant K_3 \left(\int\limits_{\widetilde{\Pi}_M} |\nabla \widetilde{W}_M|^2\mathrm{d}\xi \right)^{1/2}. \tag{74}
$$

From (67)–(74) and (64) with $\varphi = W_M + C_M$ we get

$$
\int\limits_{\Pi_M} |\nabla W_M|^2\mathrm{d}\xi \leqslant K_4,
$$

where K_4 does not depend on M.

Now we prove that $W_M + C_M$ is uniformly bounded with respect to M in $L^2(\Pi_{M'})$. It follows from the imbedding theorem that

$$
\int\limits_{\Pi_{M'}} |W_M + C_M|^2\mathrm{d}\xi
$$

$$
\leqslant C_2 \left(\int\limits_{\Pi_1} |W_M + C_M|^2\mathrm{d}\xi + \int\limits_{\Pi_{M'}} |\nabla W_M|^2\mathrm{d}\xi \right)
$$

$$
\leqslant K_5, \tag{75}
$$

where K_5 does not depend on M. Now using the diagonalization procedure we can find a sequence of M such that $W_M + C_M$ converges to $\widetilde{W}$ in $L^2(\Pi_{M_1})$ norm and ∇W_M converges weakly in $L^2(\Pi_{M_1})$ for any M_1. Let us take φ with compact support in ξ_1 and 1-periodic with respect to $\hat{\xi}$. From (64) we obtain as a limit

$$
-\int\limits_\Pi \nabla\widetilde{W}\nabla\varphi\mathrm{d}\xi \;=\; -\int\limits_\Pi f_0\varphi\mathrm{d}\xi + \int\limits_\Pi \sum_{j=1}^n f_j \frac{\partial\varphi}{\partial\xi_j}\mathrm{d}\xi + \int\limits_\Pi \nabla(\psi V)\nabla\varphi\mathrm{d}\xi
$$

$$+ \int\limits_{\gamma} [f_1] \varphi d\hat{\xi} + \int\limits_{\gamma} (W + J) \varphi d\hat{\xi}. \tag{76}$$

It can be shown that $\widetilde{W}(\xi) \to C_+$ as $\xi_1 \to \infty$ and $\widetilde{W}(\xi) \to C_-$ as $\xi_1 \to -\infty$ and

$$|\widetilde{W}(\xi) - \overline{C}| \leqslant C' e^{-\alpha|\xi_1|}, C_+, C_- = const,$$

where $\overline{C} = C_-$ for $\xi_1 < 0$ and $\overline{C} = C_+$ for $\xi_1 > 0$, C' and α are positive constants. We take $N(\xi) = \widetilde{W}(\xi) - \overline{C} + V\psi$.

The proof of the above inequality follows the derivation of a similar estimate for $\tilde{w}(\xi)$ with $\xi_1 < 0$ in the proof of Theorem 1 in section 1.

Lemma 1. *For functions* $N_{i_1}(\xi)$ *the following equality is valid:*

$$\int\limits_{\gamma} \left(\frac{\partial N_{i_1}(\xi)}{\partial \xi_1} + \delta_{i_1 1} \right) d\hat{\xi} = \int\limits_{Y} \left(\frac{\partial N_{i_1}(\xi)}{\partial \xi_1} + \delta_{i_1 1} \right) d\xi$$

$$= |Y| h_{i_1 1}^+, \tag{77}$$

where $|Y| = \int\limits_{Y} d\xi$, δ_{ks} *is the Kronecker symbol,* $h_{i_1 1}^+$ *is the constant defined by (52).*

Proof. The equation (20) for $N_{i_1}(\xi)$ and the boundary condition (21) can be written in the form

$$\Delta(N_{i_1}(\xi) + \xi_{i_1}) = 0 \text{ in } Y, \tag{78}$$

$$\frac{\partial}{\partial \nu}(N_{i_1}(\xi) + \xi_{i_1}) = 0 \text{ on } S_0. \tag{79}$$

We integrate (78) over Π_t^+, $0 < t < 1$. For $i_1 = 1$ we obtain

$$\int\limits_{\gamma} \left(\frac{\partial N_{i_1}(\xi)}{\partial \xi_1} + 1 \right) d\hat{\xi} = \int\limits_{\gamma_t} \left(\frac{\partial N_{i_1}(\xi)}{\partial \xi_1} + 1 \right) d\hat{\xi}, \tag{80}$$

where $\gamma_t = \Pi \cap \{\xi : \xi_1 = t\}$. For $i_1 = 2, \dots, n$ we get

$$\int\limits_{\gamma} \frac{\partial N_{i_1}(\xi)}{\partial \xi_1} d\hat{\xi} = \int\limits_{\gamma_t} \frac{\partial N_{i_1}(\xi)}{\partial \xi_1} d\hat{\xi}. \tag{81}$$

Integrating (80) and (81) over t from 0 to 1 and taking into account (52), we obtain (77). Lemma 1 is proved.

Thus we can solve problems for N_i, N_i^+, N_i^- and find the values for the constants:

$$h_0^{\pm} = 0, h_{i_1}^{\pm} = 0, \tag{82}$$

$$h_{i_1 i_2}^+ = \frac{1}{|Y|} \int\limits_Y \left(\frac{\partial N_{i_1}(\xi)}{\partial \xi_{i_2}} + \delta_{i_1 i_2} \right) d\xi, h_{i_1 i_2}^- = \delta_{i_1 i_2}, \tag{83}$$

$$g_0^+ = 1, g_0^- = -1, \tag{84}$$

$$j_0^+ = 0, j_0^- = 0, \tag{85}$$

$$j_{i_1}^+ = - \int\limits_\gamma \left(\frac{\partial N_{i_1}(\xi)}{\partial \xi_1} + \delta_{i_1 1} \right) d\xi = -|Y| h_{i_1 1}^+, j_{i_1}^- = \delta_{i_1 1}. \tag{86}$$

It is proved in [5], [1] that $h_{i_1 i_2}^+ = h_{i_2 i_1}^+$. Taking into account (6), (8), (9) and these values of $h_i^{\pm}$, $j_i^{\pm}$, $g_i^{\pm}$, we define

$$\Delta v_\epsilon - f(x) = 0, x \in \Omega^-, \tag{87}$$

$$\sum_{i_1, i_2 = 1}^n h_{i_1 i_2}^+ D^{i_1 i_2} v_\epsilon - f(x) = \epsilon F_1(\epsilon, x), x \in \Omega^+, \tag{88}$$

$$[v_\epsilon] = \epsilon F_2(\epsilon, x)|_{x_1 = 0}, x \in \omega, \tag{89}$$

$$\left. \frac{\partial v_\epsilon}{\partial x_1} \right|_{x_1 = -0} - |Y| \sum_{i_2 = 1}^n h_{1 i_2}^+ D^{i_2} v_\epsilon|_{x_1 = +0} = \epsilon F_3(\epsilon, x)|_{x_1 = 0}, x \in \omega. \tag{90}$$

Here F_1, F_2, F_3 are some functions which depend on

$$v_\epsilon = v_0(x) + \epsilon v_1(x) + \epsilon^2 v_2(x) + \cdots. \tag{91}$$

It is easy to see that from (87)–(90) we obtain

$$\Delta v_0(x) = f \text{ in } \Omega^-, \tag{92}$$

$$\sum_{i_1, i_2 = 1}^n h_{i_1 i_2}^+ D^{i_1 i_2} v_0 = f(x) \text{ in } \Omega^+, \tag{93}$$

$$[v_0] = 0 \text{ on } \omega, \tag{94}$$

$$\left. \frac{\partial v_0}{\partial x_1} \right|_{x_1 = -0} - |Y| \sum_{i_2 = 1}^n h_{1 i_2}^+ \left. \frac{\partial v_0}{\partial x_{i_2}} \right|_{x_1 = +0} = 0 \text{ on } \omega, \tag{95}$$

$$v_0 = 0 \text{ on } \partial\Omega. \tag{96}$$

For $k \geq 1$ we have

$$\Delta v_k(x) = \phi_k^1(v_0, \ldots, v_{k-1}), x \in \Omega^-, \tag{97}$$

$$\sum_{i_1,i_2=1}^{n} h_{i_1 i_2}^+ D^{i_1 i_2} v_k = \phi_k^2(v_0, \ldots, v_{k-1}), x \in \Omega^+, \tag{98}$$

$$[v_k] = \phi_k^3(v_0, \ldots, v_{k-1}), x \in \omega, \tag{99}$$

$$\frac{\partial v_k}{\partial x_1}\Big|_{x_1=-0} - |Y| \sum_{i_2=1}^{n} h_{1 i_2}^+ \frac{\partial v_k}{\partial x_{i_2}}\Big|_{x_1=+0} = \phi_k^4(v_0, \ldots, v_{k-1}),$$

$$x \in \omega, \tag{100}$$

$$v_k = 0 \text{ on } \partial\Omega, \tag{101}$$

where ϕ_k^1, ϕ_k^2, ϕ_k^3, ϕ_k^4 are defined from equations which we get on equating to zero coefficients of ϵ^p for any p in (6)–(9). We suppose that such sufficiently smooth solutions exist.

In this way we get the first $m+1$ members of the asymptotic expansion in the form

$$u_\epsilon^m = \sum_{k=0}^{m} \epsilon^k \sum_{|i|=k} \Big(N^i(\xi) D^i v_\epsilon(x) + N_i^+(\xi) \left(D^i v_\epsilon(x) \right)\big|_{x_1=+0}$$

$$+ N_i^-(\xi) \left(D^i v_\epsilon(x) \right)\big|_{x_1=-0} \Big), \qquad \xi = \frac{x}{\epsilon}. \tag{102}$$

The function u_ϵ^m is a solution of the following problem:

$$\Delta u_\epsilon^m - f = \epsilon^{m-1} F_\epsilon^1(x) \text{ in } \Omega_\epsilon, \tag{103}$$

$$\frac{\partial u_\epsilon^m}{\partial \nu}\Big|_{S_\epsilon} = \epsilon^m F_\epsilon^2(x)\big|_{S_\epsilon}, \tag{104}$$

$$[u_\epsilon^m] = 0, x \in \omega, \tag{105}$$

$$\left[\frac{\partial u_\epsilon^m}{\partial x_1}\right] = \epsilon^m F_\epsilon^3(x), x \in \omega, \tag{106}$$

$$u_\epsilon^m = \epsilon F_\epsilon^4(x), x \in \Gamma_\epsilon, \tag{107}$$

where the functions F_ϵ^k, $k = 1, \ldots, 4$, are bounded with respect to ϵ. Now we estimate $u_\epsilon - u_\epsilon^m$, where u_ϵ is a solution of the problem (1)–(2). We set

$$u_\epsilon - u_\epsilon^m = W_\epsilon^{(1)} + W_\epsilon^{(2)} + W_\epsilon^{(3)}, \tag{108}$$

where $W_\epsilon^{(1)}$ is a solution of the problem

$$\Delta W_\epsilon^{(1)} = 0 \text{ in } \Omega_\epsilon, \tag{109}$$

$$\frac{\partial W_\epsilon^{(1)}}{\partial \nu}\Big|_{S_\epsilon} = 0, \tag{110}$$

$$W_\epsilon^{(1)} = -u_\epsilon^m \text{ on } \Gamma_\epsilon. \tag{111}$$

The function $W_\epsilon^{(2)}$ is a solution of the problem

$$\Delta W_\epsilon^{(2)} = -\epsilon^{m-1} F_\epsilon^1(x) \text{ in } \Omega_\epsilon, \tag{112}$$

$$\left.\frac{\partial W_\epsilon^{(2)}}{\partial \nu}\right|_{S_\epsilon} = -\epsilon^m F_\epsilon^2(x), \tag{113}$$

$$W_\epsilon^{(2)} = 0 \text{ on } \Gamma_\epsilon. \tag{114}$$

For $W_\epsilon^{(3)}$ we have

$$\Delta W_\epsilon^{(3)} = 0 \text{ in } \Omega^-, \Delta W_\epsilon^{(3)} = 0 \text{ in } \Omega_\epsilon^+, \tag{115}$$

$$\left.\frac{\partial W_\epsilon^{(3)}}{\partial \nu}\right|_{S_\epsilon} = 0, \left.W_\epsilon^{(3)}\right|_{\Gamma_\epsilon} = 0, \tag{116}$$

$$\left[W_\epsilon^{(3)}\right] = 0, x \in \omega, \tag{117}$$

$$\left[\frac{\partial W_\epsilon^{(3)}}{\partial x_1}\right] = -\epsilon^m F_\epsilon^3(x), x \in \omega. \tag{118}$$

In order to estimate $u_\epsilon - u_\epsilon^m$ we estimate $W_\epsilon^{(1)}$, $W_\epsilon^{(2)}$, $W_\epsilon^{(3)}$.

Let us estimate $W_\epsilon^{(2)}(x)$. We multiply the equation (112) by $W_\epsilon^{(2)}$, integrate it over Ω_ϵ and transform the first integral by integration by parts. We get

$$-\int_{\Omega_\epsilon} \left|\nabla W_\epsilon^{(2)}\right|^2 dx$$

$$= -\epsilon^{m-1} \int_{\Omega_\epsilon} F_\epsilon^{(1)}(x) W_\epsilon^{(2)} dx + \epsilon^m \int_{S_\epsilon} F_\epsilon^{(2)}(x) W_\epsilon^{(2)} dS_\epsilon. \tag{119}$$

For the domain Ω_ϵ and $W_\epsilon^{(2)} = 0$ on Γ_ϵ the Friedrichs inequality is valid (see [1]):

$$\int_{\Omega_\epsilon} |W_\epsilon^{(2)}|^2 dx \leqslant C_0 \int_{\Omega_\epsilon} |\nabla W_\epsilon^{(2)}|^2 dx, \tag{120}$$

where the constant C_0 does not depend on ϵ. Using the imbedding theorem for Y and the transformation $\xi = \frac{x}{\epsilon}$, we obtain the inequality

$$\int_{S_\epsilon} |W_\epsilon^{(2)}|^2 dS_\epsilon \leqslant C_1 \epsilon^{-1} \int_{\Omega_\epsilon} \left(|\nabla W_\epsilon^{(2)}|^2 + |W_\epsilon^{(2)}|^2\right) dx, \tag{121}$$

where C_1 does not depend on ϵ.

From (119) and (120), (121) we have the estimate

$$\int_{\Omega_\epsilon} |W_\epsilon^{(2)}|^2 \, \mathrm{d}x + \int_{\Omega_\epsilon} |\nabla W_\epsilon^{(2)}|^2 \, \mathrm{d}x \leqslant C_2 \epsilon^{2(m-1)}, \quad C_2 = const. \qquad (122)$$

Now we estimate $W_\epsilon^{(3)}$. We have $\Delta W_\epsilon^{(3)} = 0$ in Ω^- and $\Delta W_\epsilon^{(3)} = 0$ in Ω_ϵ^+. Multiplying these equations by $W_\epsilon^{(3)}$ and integrating over Ω^- and Ω_ϵ^+, we get

$$-\int_{\Omega^-} \left|\nabla W_\epsilon^{(3)}\right|^2 \, \mathrm{d}x + \int_\omega \left(\frac{\partial W_\epsilon^{(3)}}{\partial x_1} W_\epsilon^{(3)}\right)\bigg|_{x_1=-0} \mathrm{d}\hat{x} = 0, \qquad (123)$$

$$-\int_{\Omega_\epsilon^+} \left|\nabla W_\epsilon^{(3)}\right|^2 \, \mathrm{d}x - \int_\omega \left(\frac{\partial W_\epsilon^{(3)}}{\partial x_1} W_\epsilon^{(3)}\right)\bigg|_{x_1=+0} \mathrm{d}\hat{x} = 0, \qquad (124)$$

where $\hat{x} = (x_2, \ldots, x_n)$. From (123) and (124) it follows that

$$-\int_{\Omega^- \cup \Omega_\epsilon^+} |\nabla W_\epsilon^{(3)}|^2 \mathrm{d}x$$

$$= \int_\omega \left[\frac{\partial W_\epsilon^{(3)}}{\partial x_1}\right] W_\epsilon^{(3)}\bigg|_{x_1=+0} \mathrm{d}\hat{x} + \int_\omega \left[W_\epsilon^{(3)}\right] \frac{\partial W_\epsilon^{(3)}}{\partial x_1}\bigg|_{x_1=-0} \mathrm{d}\hat{x}. (125)$$

The second integral on the right hand side of (125) is equal to zero, since $\left[W_\epsilon^{(3)}\right] = 0$. In order to estimate the first term in the right hand side of (125) we need to estimate

$$\int_\omega |W_\epsilon^{(3)}|^2 \bigg|_{x_1=+0} \mathrm{d}\hat{x}.$$

Since $\left[W_\epsilon^{(3)}\right] = 0$, we have according to the imbedding theorem and the Friedrichs inequality

$$\int_\omega |W_\epsilon^{(3)}|^2 \bigg|_{x_1=+0} \mathrm{d}\hat{x}$$

$$= \int_\omega |W_\epsilon^{(3)}|^2 \bigg|_{x_1=-0} \mathrm{d}\hat{x}$$

$$\leqslant C_1 \left(\int\limits_{\Omega^-} |W_\epsilon^{(3)}|^2 \mathrm{d}x + \int\limits_{\Omega^-} |\nabla W_\epsilon^{(3)}|^2 \mathrm{d}x \right)$$

$$\leqslant C_2 \int\limits_{\Omega^-} |\nabla W_\epsilon^{(3)}|^2 \mathrm{d}x. \tag{126}$$

Using (126) and the inequality $2ab \leqslant \frac{a^2}{d} + db^2$, we obtain from (125) that

$$\int\limits_{\Omega^- \cup \Omega_\epsilon^+} |\nabla W_\epsilon^{(3)}|^2 \mathrm{d}x \leqslant C_3 \epsilon^{2m}, \tag{127}$$

where C_3 does not depend on ϵ. From the Friedrichs inequality in Ω^- and Ω_ϵ^+ (see [1]) we obtain

$$\int\limits_{\Omega^- \cup \Omega_\epsilon^+} |W_\epsilon^{(3)}|^2 \mathrm{d}x \leqslant C_4 \epsilon^{2m}. \tag{128}$$

Consider $W_\epsilon^{(1)}$. By definition we have

$$W_\epsilon^{(1)}(x) = u_\epsilon - u_\epsilon^m - W_\epsilon^{(2)} - W_\epsilon^{(3)}. \tag{129}$$

The function $W_\epsilon^{(1)}$ belongs to $H^1(\Omega_\epsilon)$, since it is a harmonic function in $\Omega_\epsilon^+ \cup \Omega^-$ and it and its first derivatives belong to $L^2(\Omega_\epsilon^+)$ and $L^2(\Omega^-)$, because all functions in the right hand side of (129) have this property and $[W_\epsilon^{(1)}] = 0$.

The function $W_\epsilon(x) = -u_\epsilon^m - W_\epsilon^{(3)} + v_0$ satisfies the conditions $[W_\epsilon] = 0$, $W_\epsilon|_{\Gamma_\epsilon} = W_\epsilon^{(1)}|_{\Gamma_\epsilon}$. It means that W_ϵ belongs to $H^1(\Omega_\epsilon)$. According to the variational principle we have

$$\int\limits_{\Omega_\epsilon} \left| \nabla W_\epsilon^{(1)} \right|^2 \mathrm{d}x \leqslant \int\limits_{\Omega_\epsilon} |\nabla (W_\epsilon \varphi_\epsilon)|^2 \mathrm{d}x, \tag{130}$$

where φ_ϵ is a function such that $\varphi_\epsilon(x) = 1$ if $\rho(x, \partial\Omega) < \epsilon$, $\varphi_\epsilon(x) = 0$ if $\rho(x, \partial\Omega) > 2\epsilon$, $0 \leqslant \varphi_\epsilon(x) \leqslant 1$, $\varphi_\epsilon(x) \in C^\infty(\mathbf{R}^n)$, $|\nabla \varphi_\epsilon| \leqslant K\epsilon^{-1}$ with the constant K not depending on ϵ. Let us estimate the right hand side of (130).

We have

$$\int\limits_{\Omega_\epsilon} |\nabla (W_\epsilon \varphi_\epsilon)|^2 \mathrm{d}x \leqslant 2 \int\limits_{\Omega_\epsilon} |\nabla W_\epsilon \varphi_\epsilon|^2 \mathrm{d}x + 2 \int\limits_{\Omega_\epsilon} |W_\epsilon \nabla \varphi_\epsilon|^2 \mathrm{d}x. \tag{131}$$

By definition, $|\nabla(u_\epsilon^m - v_0)|$ is bounded in Ω_ϵ with respect to ϵ. For $\nabla W_\epsilon^{(3)}$ we have the estimate (127). Therefore,

$$\int\limits_{\Omega_\epsilon} |\nabla W_\epsilon \varphi_\epsilon|)^2 dx$$

$$\leqslant 2 \int\limits_{\Omega_\epsilon} \varphi_\epsilon^2 |\nabla(u_\epsilon^m - v_0)|^2 dx + 2 \int\limits_{\Omega_\epsilon} \varphi_\epsilon^2 |\nabla W_\epsilon^{(3)}|^2 dx$$

$$\leqslant C(\epsilon + \epsilon^{2m}). \tag{132}$$

For $W_\epsilon^{(3)}$ the estimate (128) is valid and $|u_\epsilon^m - v_0| \leqslant C_1 \epsilon$. Taking it into account, we obtain

$$\int\limits_{\Omega_\epsilon} |W_\epsilon \nabla \varphi_\epsilon|)^2 dx$$

$$\leqslant 2 \int\limits_{\Omega_\epsilon} (u_\epsilon^m - v_0)^2 (\nabla \varphi_\epsilon)^2 dx + 2 \int\limits_{\Omega_\epsilon} |\nabla \varphi_\epsilon W_\epsilon^{(3)}|^2 dx$$

$$\leqslant C_2(\epsilon + \epsilon^{2m-2}), \tag{133}$$

where the constants C, C_1, C_2 do not depend on ϵ. From (130), (131), (132), (133) it follows that

$$\int\limits_{\Omega_\epsilon} |\nabla W_\epsilon^{(1)}(x)|^2 dx \leqslant C_3(\epsilon + \epsilon^{2m-2}). \tag{134}$$

For the harmonic function $W_\epsilon^{(1)}$ the maximum principle is valid [6]. Therefore we have $|W_\epsilon^{(1)}(x)| \leqslant C_4 \epsilon$ and

$$\int\limits_{\Omega_\epsilon} |W_\epsilon^{(1)}(x)|^2 dx \leqslant C_5 \epsilon^2. \tag{135}$$

From the estimates for $W_\epsilon^{(2)}$, $W_\epsilon^{(3)}$, $W_\epsilon^{(1)}$, obtained above, we get

$$\int\limits_{\Omega - \cup \Omega_\epsilon^+} (u_\epsilon - u_\epsilon^m)^2 dx \leqslant C_6 \epsilon^2, \quad \int\limits_{\Omega - \cup \Omega_\epsilon^+} |\nabla(u_\epsilon - u_\epsilon^m)|^2 dx \leqslant C_7 \epsilon, \tag{136}$$

where the constants C_j do not depend on ϵ and $m \geqslant 2$.

The estimate (136) gives us the proof of the following theorem.

Theorem 2. *Let u_ϵ be a solution of the problem (1),(2). Then*

$$\|u_\epsilon - v_0\|^2_{L^2(\Omega_\epsilon)} \leqslant C\epsilon^2,$$

$$\|u_\epsilon - v_0\|^2_{H^1(\Omega^-)} + \left\|u_\epsilon - v_0 - \epsilon \sum_{j=1}^n N_j\left(\frac{x}{\epsilon}\right)\frac{\partial v_0}{\partial x_j}\right\|^2_{H^1(\Omega^+_\epsilon)} \leqslant C\epsilon,$$

where the constant C does not depend on ϵ, the function v_0 is a solution of the problem (92)–(96), $N_j(\xi)$, $j = 1,\ldots,n$, are solutions of the problem (20)–(22).

Let us consider the spectral problem corresponding to problems (1)–(2) and study the behavior of their eigenvalues and eigenfunctions as $\epsilon \to 0$.

We have proved above, as it is easy to see, that for solutions u_ϵ of the problem (1)–(2) and the solution v_0 of the problem (92)–(96) the following estimate is valid:

$$\|u_\epsilon - v_0\|_{L^2(\Omega_\epsilon)} \leqslant \epsilon C_1 \|f\|_{C^k(\overline{\Omega})}, \tag{137}$$

where C_1 is a constant which does not depend on ϵ and f, $C^k(\overline{\Omega})$ is the space of functions with continuous derivatives in $\overline{\Omega}$ up to order k.

Let $f(x) \in L^2(\Omega)$. Then there exists a sequence of functions $f_\delta \in C^\infty(\overline{\Omega})$ such that

$$\|f - f_\delta\|_{L^2(\Omega)} \leqslant h(\delta), \|D^i f_\delta\|_{L^2(\Omega)} \leqslant M_i \delta^{-|i|}, |i| = 1, 2, \ldots, \tag{138}$$

where $h(\delta) \to 0$ as $\delta \to 0$, constants M_i do not depend on δ and they depend on $\|f\|_{L^2(\Omega)}$. (For f_δ we can take averaged functions (see [2]).)

We denote by $u_{\epsilon,\delta}(x)$ a solution of the problem (1)–(2) with $f(x) = f_\delta(x)$, by v_δ a solution of the problem (92)–(96) with $f(x) = f_\delta(x)$. Then according to well-known a priori estimates

$$\|u_{\epsilon,\delta} - u_\epsilon\|_{L^2(\Omega_\epsilon)} \leqslant C_2 h(\delta), \tag{139}$$
$$\|v_\delta - v_0\|_{L^2(\Omega)} \leqslant C_3 h(\delta), \tag{140}$$

where the constants C_2, C_3 do not depend on ϵ or δ.

From (137)–(140) we have

$$\|u_{\epsilon,\delta} - v_\delta\|_{L^2(\Omega_\epsilon)} \leqslant \epsilon C_4 \delta^{-k},$$
$$\|u_\epsilon - v_0\|_{L^2(\Omega_\epsilon)} \leqslant \epsilon C_4 \delta^{-k} + (C_2 + C_3)h(\delta).$$

If we take $\delta = \epsilon^{1/k-\alpha}$, $\alpha = const$, $0 < \alpha < \frac{1}{k}$, we obtain

$$\|u_\epsilon - v_0\|_{L^2(\Omega_\epsilon)} \leqslant C_4 \epsilon^{\alpha k} + (C_2 + C_3) h(\epsilon^{1/k-\alpha}).$$

This means that for $f \in L^2(\Omega)$

$$\|u_\epsilon - v_0\|_{L^2(\Omega_\epsilon)} \to 0 \text{ as } \epsilon \to 0. \tag{141}$$

Our study of the spectrum of the problem (1),(2) is based on the application of Theorem 1.4, Ch. 3 from [1].

We introduce the space $H_\epsilon = L^2(\Omega_\epsilon)$ with the scalar product

$$(u,v)_{H_\epsilon} = \int_{\Omega_\epsilon} uv \mathrm{d}x$$

and the space H_0 with the scalar product

$$(u,v)_{H_0} = \int_{\Omega} \chi uv \mathrm{d}x,$$

where $\chi(x) = 1$ in Ω^- and $\chi(x) = |Y|$ in Ω^+.

Let us introduce operators $A_\epsilon : H_\epsilon \to H_\epsilon$, setting $A_\epsilon f = -u_\epsilon$, where u_ϵ is a solution of the problem (1),(2). Denote by A_0 the operator mapping $f \in H_0$ into $-v_0$, where v_0 is a weak solution of the problem (92)–(96).

The function v_0 is called a weak solution of the problem (92)–(96), if $v_0 \in H_0^1(\Omega)$ and for any function $\rho \in H_0^1(\Omega)$ it satisfies the identity $[v_0, \rho] = -(f, \rho)_{H_0}$, where

$$[v_0, \rho] = \int_{\Omega} \sum_{i,j=1}^{n} \hat{h}_{ij} \frac{\partial v_0}{\partial x_j} \frac{\partial \rho}{\partial x_i} \mathrm{d}x, \quad \hat{h}_{ij} = \left\{ \begin{array}{ll} \delta_{ij}, & x \in \Omega^-, \\ |Y| h_{ij}^+, & x \in \Omega^+. \end{array} \right.$$

Let us prove that for H_ϵ, H_0, A_ϵ and A_0 the conditions C_1–C_4 of Theorem 1.4, Ch. 3 from [1] are satisfied. It is easy to see that A_ϵ and A_0 are positive and compact operators. It is evident that A_ϵ is selfadjoint. If u and v are solutions of the problem (92)–(96) with $-f$ and $-g$ in the right hand side of (92), then $[u, v] = (f, v)_{H_0}$, $[u, v] = (g, u)_{H_0}$. This means that $(f, Ag)_{H_0} = (g, Af)_{H_0}$.

Therefore, conditions C_1, C_2 are satisfied.

The condition C_3 is valid because (141). In order to prove that condition C_4 is also satisfied, we use Theorem 4.3, Ch. 1 from [1] on the continuation of functions in Ω_ϵ and the Rellich–Sobolev compactness theorem.

Let us consider the eigenvalue problems for A_ϵ and A_0:

$$\begin{cases} A_\epsilon u_\epsilon^k = \mu_\epsilon^k u_\epsilon^k, k = 1, 2, \ldots, u_\epsilon^k \in H_\epsilon, \\ \mu_\epsilon^1 \geqslant \mu_\epsilon^2 \geqslant \cdots \geqslant \mu_\epsilon^k \geqslant \ldots, \mu_\epsilon^k > 0, \\ (u_\epsilon^l, u_\epsilon^j)_{H_\epsilon} = \delta_{lj}, l, j = 1, 2, \ldots, \end{cases} \tag{142}$$

$$\begin{cases} A_0 u_0^k = \mu_0^k u_0^k, k = 1, 2, \ldots, u_0^k \in H_0, \\ \mu_0^1 \geqslant \mu_0^2 \geqslant \cdots \geqslant \mu_0^k \geqslant \ldots, \mu_0^k > 0, \\ (u_0^l, u_0^j)_{H_0} = \delta_{lj}, l, j = 1, 2, \ldots, \end{cases} \tag{143}$$

where δ_{lj} is Kronecker symbol, eigenvalues μ_ϵ^k and μ_0^k, $k = 1, 2, \ldots$, form decreasing sequences and each eigenvalues is counted as many times as its multiplicity.

According to Theorem 1.4, Ch. 3 from [1] we have

$$|\mu_\epsilon^k - \mu_0^k| < C \sup \|(A_\epsilon R_\epsilon - R_\epsilon A_0)u\|_{L^2(\Omega_\epsilon)}, u \in M_k, \|u\|_{L^2(\Omega)} = 1, \tag{144}$$

where M_k is the eigenspace of the operator A_0 corresponding to the eigenvalue μ_0^k, $\epsilon \leqslant \epsilon_0(k) = const > 0$, $R_\epsilon : H_0 \to H_\epsilon$ is the restriction operator: $R_\epsilon v = v^\epsilon$, $v^\epsilon = v$ on Ω_ϵ, $v \in L^2(\Omega)$. Consider the following eigenvalue problems:

$$\begin{cases} \Delta u_\epsilon^k + \lambda_\epsilon^k u_\epsilon^k = 0 \text{ in } \Omega_\epsilon, \\ u_\epsilon^k\big|_{\Gamma_\epsilon} = 0, \dfrac{\partial u_\epsilon}{\partial \nu}\bigg|_{S_\epsilon} = 0, \\ \displaystyle\int_{\Omega_\epsilon} u_\epsilon^p u_\epsilon^j \mathrm{d}x = \delta_{pj}, p, j = 1, 2, \ldots, \\ 0 \leqslant \lambda_\epsilon^1 \leqslant \lambda_\epsilon^2 \leqslant \cdots \leqslant \lambda_\epsilon^k \leqslant \cdots; \end{cases} \tag{145}$$

$$\begin{cases} \Delta u^k + \lambda^k u^k = 0 \text{ in } \Omega^-, \displaystyle\sum_{i,j=1}^n h_{ij}^+ \dfrac{\partial^2 u^k}{\partial x_i \partial x_j} + \lambda^k u^k = 0 \text{ in } \Omega^+, \\ u^k = 0 \text{ on } \partial\Omega, [u^k] = 0, \\ \dfrac{\partial u^k}{\partial x_1}\bigg|_{x_1=-0} - |Y| \displaystyle\sum_{j=1}^n h_{1j}^+ \dfrac{\partial u^k}{\partial x_j}\bigg|_{x_1=+0} = 0, x \in \omega, \\ \displaystyle\int_\Omega \chi u^p u^j \mathrm{d}x = \delta_{pj}, p, j = 1, 2, \ldots, \\ 0 \leqslant \lambda^1 \leqslant \lambda^2 \leqslant \cdots \leqslant \lambda^k \leqslant \cdots, \end{cases} \tag{146}$$

where the eigenvalues of the problems (145) and (146) form increasing sequences and each eigenvalue is repeated as many times as its multiplicity.

It is easy to see that the eigenvalues of the problems (142),(143) and the problems (145),(146) are related by the formulas

$$\mu_\epsilon^k = (\lambda_\epsilon^k)^{-1}, \mu_0^k = (\lambda^k)^{-1}. \tag{147}$$

From (144) and the relations (147) we have

Theorem 3. *For eigenvalues λ^k of the problem (146) and eigenvalues λ_ϵ^k of the problem (145) the estimate*

$$|\lambda_\epsilon^k - \lambda^k| \leqslant C|\lambda_\epsilon^k||\lambda^k|\epsilon \tag{148}$$

is valid, where the constant C does not depend on ϵ and k; $\epsilon < \epsilon_0(k)$, $\epsilon_0(k) = const > 0$.

Proof. The inequality (148) follows from (147) and the estimate (144), since the right hand side of (144) is bounded by

$$C_1 \sup_{u \in M_k} \|\overline{u}_\epsilon - \overline{u}\|_{L^2(\Omega_\epsilon)},$$

where $\overline{u}_\epsilon$ is a solution of the problem (1),(2) with $f = -u$, and $\overline{u}$ is a solution of the problem (92)–(96) with $f = -u$, $u \in M_k$. According to Theorem 2 $\|\overline{u}_\epsilon - \overline{u}\|_{L^2(\Omega_\epsilon)}$ can be estimated by $C_2\epsilon$; C_1, C_2 do not depend on ϵ. This completes the proof of the theorem.

As a consequence of Theorem 1.7, Ch. 3 from [1] one can get estimates for the eigenfunction of the problems (145) and (146).

Remark 1. In the same way as in Theorem 2, one can consider the problem

$$L_\epsilon(u_\epsilon) = f \text{ in } \Omega_\epsilon, u_\epsilon = 0 \text{ on } \Gamma_\epsilon,$$
$$\frac{\partial u_\epsilon}{\partial \mu} = 0 \text{ on } S_\epsilon, u_\epsilon \in H_1(\Omega_\epsilon),$$

where

$$L_\epsilon = L_\epsilon^+ \text{ in } \Omega_\epsilon^+, L_\epsilon = L_\epsilon^- \text{ in } \Omega_\epsilon^-,$$
$$L_\epsilon^+ = \sum_{k,s=1}^n \frac{\partial}{\partial x_k}\left(a_{ks}^+\left(\frac{x}{\epsilon}\right)\frac{\partial}{\partial x_s}\right),$$
$$L_\epsilon^- = \sum_{k,s=1}^n \frac{\partial}{\partial x_k}\left(a_{ks}^-\left(\frac{x}{\epsilon}\right)\frac{\partial}{\partial x_s}\right),$$
$$\frac{\partial}{\partial \mu} = \sum_{k,s=1}^n a_{ks}^+\left(\frac{x}{\epsilon}\right)\nu_k\frac{\partial}{\partial x_s},$$

and L_ϵ^+ and L_ϵ^- are elliptic operators with smooth periodic coefficients in $\overline{\Omega_\epsilon^+}$ and $\overline{\Omega^-}$ correspondingly. In this case a theorem similar to Theorem 2 can be proved.

Similar results can be proved for the elasticity system. In the case of periodic boundary conditions one can get the full asymptotic expansion for u_ϵ (see [1]).

For spectral problems corresponding to problems formulated above, we can prove theorems similar to Theorem 3 based on Theorem 1.4, Ch. 3 from [1].

The results of this section were obtained jointly with W. Jäger and A.S. Shamaev.

References

[1] O.A. Oleinik, A.S. Shamaev, G.A. Yosifian, Mathematical problems in elasticity and homogenization, North-Holland, Amsterdam, 1992.

[2] S.L. Sobolev, Some applications of functional analysis in mathematical physics, Nauka, Moscow, 1988, AMS translation 1991.

[3] O.A. Oleinik, Boundary value problems for linear elliptic and parabolic equations with discontinuous coefficient, *Izvestia Akad. Nauk SSSR* ser. mat., **26**, No 1, 1961, 3–20.

[4] L. Bers, F. John, M. Schechter, Partial differential equations, Interscience Publishers, New York, 1964.

[5] A. Bensoussan, J-L. Lions, G. Papanicolau, Asymptotic analysis for periodic structures, North Holland, Amsterdam, 1978.

[6] D. Gilbarg, N.S. Trudinger, Elliptic partial differential equations of second order, 2nd edition, Springer Verlag, Berlin, 1983.

4.3   On the homogenization of solutions of the Dirichlet problem in partially perforated domains with nonperiodic structures

The problem of homogenization for elliptic equations in perforated domains were considered in many papers (see for example [1]–[4]). Here we study this problem in partially perforated domains of the general form with nonperiodic structures.

For simplicity we consider the Laplace operator. Let $\Omega, \Omega_1, \ldots, \Omega_m$ be bounded domains with smooth boundaries. Assume that $\overline{\Omega}_j \subset \Omega$ for $j = 1, \ldots, m_1$ and $\Omega_j \subset \Omega$, $\partial\Omega_j \cap \partial\Omega \neq \emptyset$ for $j = m_1+1, \ldots, m$, $\Omega_j \cap \Omega_i = \emptyset$, for $j \neq i$.

We suppose that for any positive ϵ (ϵ is a parameter) every domain Ω_j is divided in a finite number of nonintersecting domains $\Omega_{j,k}$ ($k = 1, \ldots, K(j,\epsilon)$) such that

$$\bigcup_{k=1}^{K(i,\epsilon)} \overline{\Omega}_{i,k} = \overline{\Omega}_i \quad \text{and} \quad \text{diam}\Omega_{i,k} \leqslant M\epsilon,$$

where M is a constant independent of ϵ. The domains $\Omega_{i,k}$ are called cells.

Let $\Omega_{i,k}$ contain a finite number of domains $G_{i,k,s}^{\epsilon}$ ($s = 1, \ldots, N(i,k,\epsilon)$). We set

$$\Omega_{i,k}^{\epsilon} = \Omega_{i,k} \backslash \bigcup_{s=1}^{N(i,k,\epsilon)} \overline{G}_{i,k,s}^{\epsilon},$$

$$S_{i,k}^{\epsilon} = \bigcup_{s=1}^{N(i,k,\epsilon)} (\partial G_{i,k,s}^{\epsilon} \cap \Omega), \quad S_i^{\epsilon} = \bigcup_{k=1}^{K(i,\epsilon)} S_{i,k}^{\epsilon}, \quad S_{\epsilon} = \bigcup_{i=1}^{m} S_i^{\epsilon};$$

$$\Omega_i^{\epsilon} = \Omega_i \backslash \bigcup_{k=1}^{K(i,\epsilon)} \bigcup_{s=1}^{K(i,k,\epsilon)} \overline{G}_{i,k,s}^{\epsilon}, \quad i = 1, \ldots, m,$$

$$\Gamma_i^{\epsilon} = \partial\Omega_i^{\epsilon} \backslash S_i^{\epsilon}.$$

We define the partially perforated domain Ω_{ϵ}, setting

$$\Omega_{\epsilon}^1 = \bigcup_{i=1}^{m} \Omega_i^{\epsilon}, \quad \Omega^2 = \Omega \backslash \bigcup_{i=1}^{m} \overline{\Omega}_i,$$

$$\gamma_\epsilon = \left(\bigcup_{i=1}^{m_1} \Gamma_i^\epsilon \right) \bigcup \left(\bigcup_{i=m_1+1}^{m} (\Gamma_i^\epsilon \backslash \partial \Omega) \right),$$

$$\Omega_\epsilon = \Omega_\epsilon^1 \bigcup \Omega^2 \bigcup \gamma_\epsilon, \quad \Gamma_\epsilon = \partial \Omega_\epsilon \backslash S_\epsilon.$$

As before, we denote by $H^1(\omega, \Gamma)$ the space which is the completion of the set of infinitely differentiable in $\overline{\omega}$ functions, equal to zero in a neighborhood of Γ, by the norm $H^1(\omega)$.

Assume that

(1) for any cell $\Omega_{i,k}^\epsilon$ the Friedrichs inequality is valid: if $u \in H^1(\Omega_{i,k}^\epsilon, S_{i,k}^\epsilon)$, then

$$\|u\|_{L^2(\Omega_{i,k}^\epsilon)} \leqslant K_0 \epsilon \|\nabla u\|_{L^2(\Omega_{i,k}^\epsilon)} \qquad (1)$$

where K_0 does not depend on u, ϵ, i, k; $\nabla u = \left(\frac{\partial u}{\partial x_1}, \dots, \frac{\partial u}{\partial x_n} \right)$;

(2) for the cells $\Omega_{i,k}^\epsilon$ for which $\gamma_{i,k}^\epsilon = \partial \Omega_{i,k}^\epsilon \cap (\partial \Omega_i \backslash \partial \Omega) \neq \emptyset$ the following imbedding theorem is valid: if $u \in H^1(\Omega_{i,k}^\epsilon, S_{i,k}^\epsilon)$, then

$$\|u\|_{L^2(\gamma_{i,k}^\epsilon)} \leqslant K_1 \sqrt{\epsilon} \|\nabla u\|_{L^2(\Omega_{i,k}^\epsilon)}, \qquad (2)$$

the constant K_1 does not depend on ϵ, k, u.

Remark 1. The inequality (1) is valid, for example, in the case when Ω_i^ϵ is ϵ-perforated domain with a periodic structure [3]. This inequality is also satisfied, if for some s the domain $G_{i,k,s}^\epsilon$ contains a ball of the radius $\tau_\epsilon = d\epsilon$, where the constant d does not depend on i, k, ϵ, s.

Remark 2. The inequality (2) is valid, for example, in the case when the inequality (1) is valid and every cell $\Omega_{i,k}$, such that $\partial \Omega_{i,k} \cap \partial \Omega_i \backslash \partial \Omega \neq \emptyset$, is diffeomorphic to the cube Q_ϵ with the side ϵ, and the Jacobian of this diffeomorphism is bounded from below and above by positive constants, which do not depend on i, k, ϵ, where $u(x)$ is defined as zero in $G_{i,k,s}^\epsilon$.

Let us consider the boundary-value problem

$$\Delta u_\epsilon = f \quad \text{in} \quad \Omega_\epsilon, \qquad (3)$$

$$u_\epsilon = 0 \quad \text{on} \quad \partial \Omega_\epsilon, \quad f \in L^2(\Omega), \quad u_\epsilon \in H^1(\Omega_\epsilon, \partial \Omega_\epsilon). \qquad (4)$$

Using the integral identity for the problem (3),(4) and inequality (1), one can easy to get the estimates:

$$\|u_\epsilon\|_{L^2(\Omega_\epsilon^1)} \leqslant K_2 \epsilon, \quad \|u_\epsilon\|_{L^2(\Omega^2)} \leqslant K_3,$$

$$\|\nabla u_\epsilon\|_{L^2(\Omega_\epsilon)} \leqslant K_4,$$

where K_j here and in what follows are constants independent of ϵ.

Let ν be an exterior unit normal vector to $\partial\Omega_i$. We introduce the functions $v(x)$ and $w_\epsilon(x)$ as weak solutions of the problems

$$\Delta v = f \quad \text{in} \quad \Omega^2, \quad v = 0 \quad \text{on} \quad \partial\Omega^2, \tag{5}$$

$$\begin{cases} \Delta w_\epsilon = 0 \quad \text{in} \quad \Omega^1_\epsilon \bigcup \Omega^2, \\ w_\epsilon = 0 \quad \text{on} \quad \partial\Omega_\epsilon, \\ [w_\epsilon]_{\gamma_\epsilon} = 0, \quad \left[\dfrac{\partial w}{\partial \nu}\right]_{\gamma_\epsilon} = \dfrac{\partial v}{\partial \nu}\Big|_{\substack{x \to \gamma_\epsilon, \\ x \in \Omega^2}}, \end{cases} \tag{6}$$

where

$$[\varphi]_{\gamma_\epsilon} = \varphi|_{\substack{x \to \gamma_\epsilon \\ x \in \Omega^1_\epsilon}} - \varphi|_{\substack{x \to \gamma_\epsilon, \\ x \in \Omega^2}}.$$

Such solutions can be obtained using the Lax–Milgram theorem. Since we assume that $\partial\Omega_i$ and $\partial\Omega$ intersects with an angle less than π, according to the theorem proved in [5] the function $v(x) \in H^2(\Omega^2)$.

The weak solution of the problem (6) is a function $w_\epsilon \in H^1(\Omega_\epsilon, \partial\Omega_\epsilon)$ which satisfies the integral identity

$$\int\limits_{\Omega_\epsilon} \nabla w_\epsilon \nabla \varphi dx - \int\limits_{\gamma_\epsilon} \varphi \, \dfrac{\partial v}{\partial \nu}\Big|_{\substack{x \to \gamma_\epsilon, \\ x \in \Omega^2}} ds = 0, \tag{7}$$

for any function $\varphi \in H^1(\Omega_\epsilon, \partial\Omega_\epsilon)$.

Let us derive the estimate for w_ϵ. Taking $\varphi = w_\epsilon$ in (7), we obtain

$$\int\limits_{\Omega_\epsilon} |\nabla w_\epsilon|^2 \, dx = \int\limits_{\gamma_\epsilon} w_\epsilon \, \dfrac{\partial v}{\partial \gamma}\Big|_{\substack{x \to \gamma_\epsilon, \\ x \in \Omega^2}} ds,$$

and

$$\|\nabla w_\epsilon\|^2_{L^2(\Omega_\epsilon)} \leqslant K_5\|w_\epsilon\|_{L^2(\gamma_\epsilon)}. \tag{8}$$

Let us estimate the right hand side of (8). We consider cells $\Omega^\epsilon_{i,k}$ such that $\gamma^\epsilon_{i,k} \neq \emptyset$ and denote then $X^\epsilon_j(j = 1, \ldots, M(\epsilon))$. We set

$$\theta^\epsilon_j = \overline{X}^\epsilon_j \bigcap \left(\bigcup_{i=1}^m \partial\Omega_i\right).$$

Using inequality (2) we obtain

$$
\begin{aligned}
\|w_\epsilon\|^2_{L^2(\gamma_\epsilon)} &\leqslant \sum_{j=1}^{M(\epsilon)} \|w_\epsilon\|^2_{L^2(\theta^\epsilon_j)} \\
&\leqslant K_6\epsilon \sum_{j=1}^{M(\epsilon)} \|\nabla w_\epsilon\|^2_{L^2(X^\epsilon_j)} \\
&\leqslant K_7\epsilon \|\nabla w_\epsilon\|^2_{L^2(\Omega_\epsilon)}. \tag{9}
\end{aligned}
$$

From (8) and (9) it follows that

$$
\|\nabla w_\epsilon\|_{L^2(\Omega_\epsilon)} \leqslant K_8\sqrt{\epsilon}. \tag{10}
$$

According to (1) and (10) we have

$$
\begin{aligned}
\|w_\epsilon\|^2_{L^2(\Omega^1_\epsilon)} &\leqslant \sum_{j=1}^m \|w_\epsilon\|^2_{L^2(\Omega^\epsilon_j)} = \sum_{j=1}^m \sum_{k=1}^{K(j,\epsilon)} \|w_\epsilon\|^2_{L^2(\Omega^\epsilon_{j,k})} \\
&\leqslant K_9\epsilon^2 \sum_{j=1}^m \sum_{k=1}^{K(j,\epsilon)} \|\nabla w_\epsilon\|^2_{L^2(\Omega^\epsilon_{j,k})} \leqslant K_{10}\epsilon^3.
\end{aligned}
$$

Therefore,

$$
\|w_\epsilon\|_{L^2(\Omega^1_\epsilon)} \leqslant K_{11}\epsilon\sqrt{\epsilon}. \tag{11}
$$

From the inequality (10) and the Friedrichs inequality for Ω^2 we obtain

$$
\|w_\epsilon\|_{L^2(\Omega^2)} \leqslant K_{12}\sqrt{\epsilon}. \tag{12}
$$

Thus, the following lemma is proved.

Lemma 1. *Let w_ϵ be a weak solution of the problem (6). Then for w_ϵ the estimates (10)–(12) are valid.*

We set

$$
\overline{f}(x) = f(x) \quad \text{in} \quad \bigcup_{j=1}^m \Omega_j, \quad \overline{f}(x) = 0 \quad \text{in } \Omega^2.
$$

Let v_ϵ be a solution of the problem

$$
\Delta v_\epsilon = \overline{f} \quad \text{in } \Omega_\epsilon, \quad v_\epsilon = 0 \quad \text{on } \partial\Omega_\epsilon, \quad u \in H^1(\Omega_\epsilon, \partial\Omega_\epsilon). \tag{13}
$$

From the integral identity for the problem (13) follows that

$$
\|\nabla v_\epsilon\|^2_{L^2(\Omega_\epsilon)} = -\int_{\Omega^1_\epsilon} f v_\epsilon \mathrm{d}x. \tag{14}
$$

Using (1) and (14) we obtain

$$
\begin{aligned}
\|\nabla v_\epsilon\|^2_{L^2(\Omega_\epsilon)} &\leqslant K_{13}\|v_\epsilon\|_{L^2(\Omega^1_\epsilon)} \\
&\leqslant K_{13}\left[\sum_{j=1}^{m}\sum_{k=1}^{K(j,\epsilon)}\|v_\epsilon\|^2_{L^2(\Omega^\epsilon_{j,k})}\right]^{1/2} \\
&\leqslant K_{14}\epsilon\|\nabla v_\epsilon\|_{L^2(\Omega_\epsilon)}.
\end{aligned}
$$

Therefore,

$$\|\nabla v_\epsilon\|_{L^2(\Omega_\epsilon)} \leqslant K_{14}\epsilon. \tag{15}$$

From (1) and (15) we have

$$\|v_\epsilon\|_{L^2(\Omega^1_\epsilon)} \leqslant K_{15}\epsilon^2, \quad \|v_\epsilon\|_{L^2(\Omega^2)} \leqslant K_{16}\epsilon. \tag{16}$$

Lemma 2. *Let v_ϵ be a weak solution of the problem (13). Then the estimates (15),(16) are valid.*

By virtue of the uniqueness of a weak solution of the problem (3),(4) and the definition of the functions v, w_ϵ, v_ϵ we have

$$u_\epsilon = \overline{v} + w_\epsilon + v_\epsilon,$$

where $\overline{v} = v$ in Ω^2, $\overline{v} = 0$ in $\Omega\backslash\Omega^2$.

From Lemma 1 and 2 the following theorem follows.

Theorem 1. *Let u_ϵ be a solution of the problem (3),(4), v be a solution of the problem (5). Then*

$$\|u_\epsilon\|_{L^2(\Omega^1_\epsilon)} \leqslant K_{17}\epsilon\sqrt{\epsilon}, \tag{17}$$

$$\|u_\epsilon - v\|_{L^2(\Omega^2)} \leqslant K_{18}\sqrt{\epsilon}. \tag{18}$$

Let us consider the spectral problem, corresponding to the problem (3),(4):

$$\nabla u^k_\epsilon + \lambda^k_\epsilon u_\epsilon = 0 \quad \text{in } \Omega_\epsilon, \tag{19}$$

$$u^k_\epsilon = 0 \quad \text{on } \partial\Omega_\epsilon, \tag{20}$$

$$\int_{\Omega_\epsilon} u^k_\epsilon u^p_\epsilon \mathrm{d}x = \delta_{kp}, \tag{21}$$

where δ_{kp} is the Kronecker symbol. As it is known, the eigenvalues $\{\lambda_\epsilon^k\}$ of the problem (19)–(21) form an nondecreasing sequence where every eigenvalue is repeated as many times as its multiplicity.

Consider the limit eigenvalue problem:

$$\nabla u^k + \lambda^k u^k = 0 \quad \text{in} \quad \Omega^2, \tag{22}$$

$$u^k = 0 \quad \text{on} \quad \partial\Omega^2, \tag{23}$$

$$\int_{\Omega^2} u^k u^p \, \mathrm{d}x = \delta_{kp}. \tag{24}$$

The eigenvalues $\{\lambda^k\}$ form also an nondecreasing sequence where every eigenvalue is repeated as many times as its multiplicity.

The behavior of u_ϵ^k and λ_ϵ^k as $\epsilon \to 0$ can be studied in the same way as the spectral problem is considered in section 1 on the base of Th.1.7, Ch. III from [3]. Using the estimates (17),(18) we can prove the following theorem.

Theorem 2. *Let $\{\lambda_\epsilon^k\}$ be the sequence of the eigenvalues of the problem (19)–(21), $\{\lambda^k\}$ be the sequence of the eigenvalues of the problem (22)–(24). Then*

$$\left| \frac{1}{\lambda_\epsilon^k} - \frac{1}{\lambda^k} \right| \leqslant K_{19}\sqrt{\epsilon}.$$

Let $k \geqslant 0$, $p \geqslant 1$ be integers such that

$$\frac{1}{\lambda^k} > \frac{1}{\lambda^{k+1}} = \cdots = \frac{1}{\lambda^{k+p}} > \frac{1}{\lambda^{k+p+1}}.$$

Then for any eigenfunction w corresponding to the eigenvalue λ^{k+1} there exists a linear combination $\overline{u}_\epsilon$ of the eigenfunctions $u_\epsilon^{k+1}, \ldots, u_\epsilon^{k+p}$ of the problem (19)–(21) such that

$$\|\overline{u}_\epsilon - w\|_{L^2(\Omega^2)} + \|\overline{u}_\epsilon\|_{L^2(\Omega_\epsilon^1)} \leqslant K_{20}\sqrt{\epsilon}.$$

The results of this section are obtained jointly with T.A. Shaposhnikova.

References

[1] V.A. Marchenko, E. Ya. Khruslov, Boundary value problems in domains with a fine grained boundary, Naukova Dumka, Kiev, 1974.

[2] D. Cioranescu, J. Saint Jean Paulin, Homogenization in open sets with holes, *Journ. Math. Anal. Appl.* **71**, 1979, 590–607.

[3] O.A. Oleinik, A.S. Shamaev, G.A. Yosifian, Mathematical problems of elasticity and homogenization, North Holland, Amsterdam, 1992.

[4] V.V. Jikov, S.M. Kozlov, O.A. Oleinik, Homogenization of differential operators and integral functionals, Springer Verlag, 1994.

[5] V.A. Kondratiev, On the smoothness of solutions of the Dirichlet problem for an elliptic second order equation in a piecewise smooth domain. *Differ. equations* **6**, 1970, 10, 1831–1843.

4.4 On the asymptotics of solutions and eigenvalues of an elliptic problem with rapidly alternating type of boundary conditions

Such problems are considered in several papers ([1]–[8]). This type of problems arises in many fields of physics, biology and mathematical physics. In this section we consider one of them.

Let Ω be a smooth domain in $\mathbf{R}^n$, $n \geqslant 2$, and let $\partial\Omega$ be its boundary. We suppose that $\partial\Omega = \Gamma_\epsilon \cup \gamma_\epsilon$ and consider the boundary-value problem,

$$L(u_\epsilon) \equiv \frac{\partial}{\partial x_j}\left(a_{ij}(x)\frac{\partial}{\partial x_i}\right) = f(x) \text{ in } \Omega, \tag{1}$$

$$u_\epsilon = 0 \text{ on } \gamma_\epsilon, \tag{2}$$

$$\frac{\partial u_\epsilon}{\partial \mu} + a(x)u_\epsilon = 0 \text{ on } \Gamma_\epsilon, \tag{3}$$

where $a(x) \geqslant 0$, $a(x)$, $a_{ij}(x)$ are bounded measurable functions, $\mathfrak{æ}_1|\xi|^2 \leqslant a_{ij}(x)\xi_i\xi_j \leqslant \mathfrak{æ}_2|\xi|^2$, $\mathfrak{æ}_1$, $\mathfrak{æ}_2 = const > 0$, $a_{ij}(x) = a_{ji}(x)$, $f(x) \in L_2(\Omega)$, $\frac{\partial u_\epsilon}{\partial \mu} \equiv a_{ij}(x)\frac{\partial u_\epsilon}{\partial x_i}\nu_j$, $\nu = (\nu_1,\ldots,\nu_n)$ is an outward normal vector to the boundary $\partial\Omega$, Γ_ϵ consists of the union of sets Γ_ϵ^k, $k = 1,\ldots,N_\epsilon$, with $\operatorname{diam}\Gamma_\epsilon^k \leqslant \epsilon$, and the distance between them is greater than or equal to 2ϵ, ϵ being a small positive parameter. The set γ_ϵ stands for $\partial\Omega\backslash\Gamma_\epsilon$. Here and throughout we use the usual convention of repeated indices.

We will study the limit behavior of solutions of the problem (1)–(3), when ϵ tends to zero. Existence and uniqueness of the solutions u_ϵ of the problem (1)–(3) in the space $H^1(\Omega,\gamma_\epsilon)$ can be proved, using the Lax–Milgram lemma (see, for instance, [9]). The space $H^1(\Omega,\gamma_\epsilon)$ is defined as the completion of the functions from the space $C^\infty(\overline{\Omega})$ vanishing in a neighborhood of γ_ϵ with respect to the norm

$$\|u\|_{H^1(\Omega)} \equiv \left(\int\limits_\Omega \left(u^2 + |\nabla u|^2\right)\mathrm{d}x\right)^{\frac{1}{2}}.$$

In [1]–[8] either periodic cases of distribution of Γ_ϵ^k were considered or only a weak convergence of u_ϵ in $H^1(\Omega)$ to the solution of the limit problem as $\epsilon \to 0$ was proved. In this lecture we give the estimates of the deviation in $H^1(\Omega)$ norm of the solution u_ϵ from the solution u to the limit problem.

Lemma 1. *For the functions u from the space $H^1(\Omega, \gamma_\epsilon)$ the estimate*

$$\int_{\Omega_\eta} u^2 \mathrm{d}x \leqslant C\eta^2 \int_{\Omega_\eta} |\nabla u|^2 \mathrm{d}x \qquad (4)$$

holds, where the constant C doesn't depend on ϵ, η and u, $\Omega_\eta = \{x : x \in \Omega, \rho(x, \partial\Omega) \leqslant \eta\}$, $\rho(x, \partial\Omega)$ is equal to the distance between x and $\partial\Omega$, $\epsilon \leqslant \eta$.

Proof. Let q_ϵ^k be a ball with radius ϵ and let $p_\epsilon^k \subset \Gamma_\epsilon^k$ be a centre of the ball q_ϵ^k. Also let Q_ϵ^k be a ball with radius 2ϵ with the same centre, $s_\epsilon^k = \partial\Omega \cap q_\epsilon^k$, $S_\epsilon^k = \partial\Omega \cap Q_\epsilon^k$. The function u is, obviously, equal to 0 on $S_\epsilon^k \backslash s_\epsilon^k$. The domain G_η^k, which is a union of the inward normals to the set S_ϵ^k with the length η, is considered. Since the boundary $\partial\Omega$ is smooth, the domains G_η^k are diffeomorphic for all k. Furthermore, diamG_η^k is of order η. Then the Friedrichs inequality for the domains G_η^k (see [10]) gives us the inequality (4) in G_η^k with the constant C, which doesn't depend on ϵ, η and k. Since $u = 0$ on $\partial\Omega \backslash \left(\bigcup_k S_\epsilon^k \right)$, then as usual in the domain $\Omega_\eta \backslash \left(\bigcup_k G_\eta^k \right)$ we obtain the inequality (4), using the representation of the function u as an integral of its normal derivative. The summation of these inequalities gives us the inequality (4).

Lemma 2. *For the function u from the space $H^1(\Omega, \gamma_\epsilon)$ the estimate*

$$\int_\Omega u^2 \mathrm{d}x \leqslant C_1 \int_\Omega |\nabla u|^2 \mathrm{d}x \qquad (5)$$

holds, where the constant C_1 doesn't depend on ϵ and u.

Proof. By the mean-value theorem for an integral and (4), for $\eta = \epsilon$ we obtain that there exists $\epsilon_0 \leqslant \epsilon$ such that

$$\int_{l_{\epsilon_0}} u^2 \mathrm{d}s \leqslant C\epsilon \int_{\Omega_\epsilon} |\nabla u|^2 \mathrm{d}x, \qquad (6)$$

where $l_{\epsilon_0} = \{x : x \in \Omega, \rho(x, \partial\Omega) = \epsilon_0\}$. From the imbedding theorem (see [11]), we obtain

$$\int_{\Omega \backslash \Omega_{\epsilon_0}} u^2 \mathrm{d}x \leqslant C_2 \left(\int_{l_{\epsilon_0}} u^2 \mathrm{d}s + C \int_{\Omega \backslash \Omega_{\epsilon_0}} |\nabla u|^2 \mathrm{d}x \right), \qquad (7)$$

where the constant C_2 doesn't depend on ϵ and u, because of smoothness of the boundary $\partial\Omega$. The summation of the inequalities (4) and (7) gives us the inequality (5).

Lemma 3. *The solutions u_ϵ of the problem (1)–(3) are uniformly bounded with respect to ϵ in $H^1(\Omega)$.*

Proof. The definition of the weak solution u_ϵ in $H^1(\Omega, \gamma_\epsilon)$ of the problem (1)–(3) gives us the following integral identity:

$$\int\limits_\Omega a_{ij}(x)\frac{\partial u_\epsilon}{\partial x_i}\frac{\partial v}{\partial x_j}dx + \int\limits_{\Gamma_\epsilon} a(x)u_\epsilon v ds$$
$$= -\int\limits_\Omega fv dx \qquad (8)$$

for all $v \in H^1(\Omega, \gamma_\epsilon)$. According to ellipticity of the operator L and the inequality $ab \leqslant \frac{1}{\alpha}a^2 + \alpha b^2$, we obtain that

$$\ae_1 \int\limits_\Omega |\nabla u_\epsilon|^2 dx \leqslant \frac{1}{\alpha}\int\limits_\Omega f^2(x)dx + \alpha \int\limits_\Omega u_\epsilon^2 dx.$$

Then by the Friedrichs inequality (5), we have

$$\ae_1 \int\limits_\Omega |\nabla u_\epsilon|^2 dx \leqslant \frac{1}{\alpha}\int\limits_\Omega f^2(x)dx + \alpha C_1 \int\limits_\Omega |\nabla u_\epsilon|^2 dx.$$

By setting $\alpha = \frac{\ae_1}{2C_1}$, we find that

$$\int\limits_\Omega |\nabla u_\epsilon|^2 dx \leqslant C_3, \qquad (9)$$

where the constant C_3 doesn't depend on ϵ. The uniform estimate of u_ϵ in $H^1(\Omega)$ comes out from (5) and (9).

Let $u_0(x)$ be a weak solution of the problem

$$L(u_0) = f(x) \text{ in } \Omega, \qquad (10)$$
$$u_0 = 0 \text{ on } \partial\Omega. \qquad (11)$$

We define the function $\phi(s) \in C^\infty(\mathbf{R}^1)$ such that $\phi(s) = 0$, when $s \in [-\infty, 1]$, $\phi(s) = 1$, when $s \geqslant 1 + \sigma$, $0 < \sigma < \frac{1}{2}$, $0 \leqslant \phi(s) \leqslant 1$. Let

$\phi_\epsilon^k(x) \, = \, \phi\left(\dfrac{|\ln \epsilon|}{|\ln r_k|}\right)$, where $(r_k, \theta_k^1, \cdots, \theta_k^{n-1})$ is a local system of polar coordinates, whose centre is in $p_\epsilon^k \subset \Gamma_\epsilon^k$. Let

$$\phi_\epsilon(x) = \prod_k \phi_\epsilon^k(x).$$

Theorem 1. *For the solutions u_ϵ of the problem (1)–(3) and the solution u_0 of the problem (10),(11) the estimate*

$$\int\limits_\Omega |\nabla(u_\epsilon - u_0)|^2 \phi_\epsilon^2(x) \mathrm{d}x \leqslant C_4 |\ln \epsilon|^{-\delta} \tag{12}$$

is valid, where the constant C_4 does not depend on ϵ, $0 < \delta < 2 - \frac{2}{n}$, $N_\epsilon = O\left(|\ln \epsilon|^{(1-\frac{\epsilon}{2})n - 1}\right)$ as $\epsilon \to 0$.

Proof. Subtracting the integral identity of the problem (10),(11) from the integral identity of the problem (1)–(3) and setting $v = (u_\epsilon - u_0)\phi_\epsilon^2$, we obtain

$$\int\limits_\Omega \left(a_{ij}(x) \frac{\partial(u_\epsilon - u_0)}{\partial x_i} \frac{\partial((u_\epsilon - u_0)\phi_\epsilon^2)}{\partial x_j}\right) \mathrm{d}x = 0$$

and, therefore,

$$\ae_1 \int\limits_\Omega |\nabla(u_\epsilon - u_0)|^2 \phi_\epsilon^2(x)\mathrm{d}x$$

$$\leqslant \left|2 \int\limits_\Omega \left(a_{ij}(x)\frac{\partial(u_\epsilon - u_0)}{\partial x_i}(u_\epsilon - u_0)\phi_\epsilon \frac{\partial \phi_\epsilon}{\partial x_j}\right) \mathrm{d}x\right|$$

$$\leqslant C_5 \int\limits_\Omega (u_\epsilon - u_0)^2 |\nabla \phi_\epsilon|^2 \mathrm{d}x + \frac{\ae_1}{2}\int\limits_\Omega |\nabla(u_\epsilon - u_0)|^2 \phi_\epsilon^2 \mathrm{d}x, \tag{13}$$

where C_5 does not depend on ϵ. The next inequality follows from (13):

$$\int\limits_\Omega |\nabla(u_\epsilon - u_0)|^2 \phi_\epsilon^2(x)\mathrm{d}x$$

$$\leqslant C_6 \sum_{k=1}^{N_\epsilon} \int\limits_{\mathfrak{Q}_\epsilon^k} (u_\epsilon - u_0)^2 |\nabla \phi_\epsilon^k|^2 \mathrm{d}x, \tag{14}$$

where $\mathfrak{Q}_\epsilon^k$ is a ball with radius $\epsilon^{\frac{1}{1+\sigma}}$, and centre p_ϵ^k. Note that

$$|\nabla \phi_\epsilon^k| \leqslant C_7 |\ln \epsilon| \frac{1}{\ln^2 r_k} \frac{1}{r_k}. \tag{15}$$

Let us consider next the imbedding theorem of S.L. Sobolev [11]: the space $H^1(\Omega)$ is continuously imbedded in the space $L_q(\Omega)$, if the domain Ω is representable as a finite union of star-shaped domains and $q \leqslant \frac{2n}{n-2}$.

Using this theorem, we obtain the estimate of the right part of (14).

By using the estimate (15) and the Hölder inequality, we deduce

$$\int\limits_{\mathfrak{Q}_\epsilon^k} (u_\epsilon - u_0)^2 \left(|\ln \epsilon| |\ln r_k|^{-2} r_k^{-1}\right)^2 dx$$

$$\leqslant |\ln \epsilon|^2 \left(\int\limits_{\mathfrak{Q}_\epsilon^k} (u_\epsilon - u_0)^{2p_1} dx\right)^{\frac{1}{p_1}}$$

$$\times \left(\int\limits_{\mathfrak{Q}_\epsilon^k} (|\ln r_k|^{-4} r_k^{-2})^{p_2} dx\right)^{\frac{1}{p_2}}, \tag{16}$$

where $\frac{1}{p_1} + \frac{1}{p_2} = 1$. We suppose that $2p_1 = q = \frac{2n}{n-2}$, $p_2 = \frac{n}{2}$. It is easy to see that

$$\int\limits_{\mathfrak{Q}_\epsilon^k} (|\ln r_k|^{-4} r_k^{-2})^{p_2} dx \leqslant C_8 |\ln \epsilon|^{1-2n}, \tag{17}$$

where the constant C_8 does not depend on k and ϵ.

From the inequalities (16) and (17) we obtain

$$\int\limits_{\mathfrak{Q}_\epsilon^k} (u_\epsilon - u_0)^2 |\nabla \phi_\epsilon^k|^2 dx$$

$$\leqslant C_9 |\ln \epsilon|^{\frac{2}{n} - 2} \left(\int\limits_{\mathfrak{Q}_\epsilon^k} (u_\epsilon - u_0)^{\frac{2n}{n-2}} dx\right)^{\frac{n-2}{n}}, \tag{18}$$

where the constant C_9 does not involve ϵ and k. Thus, we have

$$\int\limits_{\Omega} |\nabla(u_\epsilon - u_0)|^2 \phi_\epsilon^2(x) dx$$

$$\leqslant C_{10} \sum_{k=1}^{N_\epsilon} |\ln \epsilon|^{\frac{2}{n}-2} \left(\int_{\Omega_\epsilon^k} (u_\epsilon - u_0)^{\frac{2n}{n-2}} \mathrm{d}x \right)^{\frac{n-2}{n}}. \qquad (19)$$

Using the Hölder inequality and the imbedding theorem, we obtain

$$\sum_{k=1}^{N_\epsilon} \left(\int_{\Omega_\epsilon^k} (u_\epsilon - u_0)^{\frac{2n}{n-2}} \mathrm{d}x \right)^{\frac{n-2}{n}}$$

$$\leqslant \left(\sum_{k=1}^{N_\epsilon} 1 \right)^{\frac{2}{n}} \left(\sum_{k=1}^{N_\epsilon} \int_{\Omega_\epsilon^k} (u_\epsilon - u_0)^{\frac{2n}{n-2}} \mathrm{d}x \right)^{\frac{n-2}{n}}$$

$$\leqslant (N_\epsilon)^{\frac{2}{n}} \|u_\epsilon - u_0\|_{L_q(\Omega)}^2$$

$$\leqslant (N_\epsilon)^{\frac{2}{n}} \|u_\epsilon - u_0\|_{H^1(\Omega)}^2. \qquad (20)$$

Lemma 3 and the smoothness of the solution u_0 lead us to the conclusion that the norms $\|u_\epsilon - u_0\|_{H^1(\Omega)}^2$ are uniformly bounded with respect to ϵ. Therefore, if we take $N_\epsilon = O\left(|\ln \epsilon|^{(1-\frac{\delta}{2})n-1}\right)$ as $\epsilon \to 0$, then from (19) and (20) we obtain (12), where δ satisfies the inequality $0 < \delta < 2 - \frac{2}{n}$. Theorem 1 is proved.

Remark 1. The proof of Theorem 1 doesn't use the maximum principle, which takes place for u_ϵ when $a_{ij}(x)$, $f(x)$, $a(x)$ are smooth (see [12]–[15]). Therefore we can also prove the analogous theorem for the system of stationary linear elasticity. Under the assumption of the indicated smoothness of the data of the problem (1)–(3) and by using the maximum principle, we shall prove the following theorem.

Theorem 2. *Let $a_{ij}(x)$, $f(x)$, $a(x)$ be sufficiently smooth functions and let Γ_ϵ have a smooth boundary on $\partial\Omega$. Then, if $n \geqslant 3$ for the solution u_ϵ of the problem (1)–(3) and the solution u_0 of the problem (10),(11), the estimate*

$$\int_\Omega |\nabla(u_\epsilon - u_0)|^2 \phi_\epsilon^2(x)\mathrm{d}x \leqslant C_{11}\epsilon^\delta \qquad (21)$$

holds, where the constant C_{11} does not depend on ϵ, $0 < \delta < n - 2$ and $N_\epsilon = O\left(|\ln \epsilon|^2\epsilon^{2-n+\delta}\right)$ as $\epsilon \to 0$.

Proof. We take the function ϕ_ϵ with σ, which satisfies the inequality $\delta < \frac{n-2}{1+\sigma}$. The inequalities (14) and (15) give us the inequality

$$\int_\Omega |\nabla(u_\epsilon - u_0)|^2 \phi_\epsilon^2(x) dx$$

$$\leqslant C_6 \max_\Omega |(u_\epsilon - u_0)|^2 \sum_{k=1}^{N_\epsilon} \int_{\Omega_\epsilon^k} (|\ln \epsilon||\ln r_k|^{-2} r_k^{-1})^2 dx$$

$$\leqslant C_{12} N_\epsilon |\ln \epsilon|^{-2} \epsilon^{\frac{n-2}{1+\sigma}}.$$

Then, if we take $N_\epsilon = O\left(|\ln \epsilon|^2 \epsilon^{\frac{2-n}{1+\sigma}+\delta}\right)$ as $\epsilon \to 0$, we obtain (21). The theorem is proved.

Theorem 3. *For the solution u_ϵ of the problem (1)–(3) and the solution u_0 of the problem (10),(11) we have the estimate*

$$\int_\Omega (u_\epsilon - u_0)^2 dx \leqslant C_{13} |\ln \epsilon|^{-\delta}, \qquad 0 < \delta < 2 - \frac{2}{n}, \tag{22}$$

for $N_\epsilon = O(|\ln \epsilon|^{(1-\frac{\delta}{2})n-1})$ as $\epsilon \to 0$. If the conditions of Theorem 2 hold, then

$$\int_\Omega (u_\epsilon - u_0)^2 dx \leqslant C_{14} \epsilon^\alpha, \qquad 0 < \delta < n-2, \alpha = \min(\delta, \frac{1}{1+\sigma}), \tag{23}$$

for $N_\epsilon = O(|\ln \epsilon|^2 \epsilon^{2-n+\delta})$ as $\epsilon \to 0$, where C_{13}, C_{14} don't depend on ϵ.

Proof. By Lemmas 1 and 3,

$$\int_{\Omega_\beta} u_\epsilon^2 dx \leqslant C_{15} \beta^2, \quad \beta = \epsilon^{\frac{1}{1+\sigma}}. \tag{24}$$

Since $u_0 = 0$ on $\partial\Omega$,

$$\int_{\Omega_\beta} u_0^2 dx \leqslant C_{16} \beta^2. \tag{25}$$

Therefore, the mean-value theorem for integrals gives us the conclusion, that there exists ϵ_0 such that $\epsilon_0 \leqslant \beta$, and

$$\int_{l_{\epsilon_0}} (u_\epsilon - u_0)^2 ds \leqslant C_{17} \beta. \tag{26}$$

It is easy to see that

$$\int\limits_{l_\beta} (u_\epsilon - u_0)^2 \mathrm{d}s$$

$$\leqslant C_{18} \left(\int\limits_{l_{\epsilon_0}} (u_\epsilon - u_0)^2 \mathrm{d}s + \beta \int\limits_{\Omega_\beta} |\nabla(u_\epsilon - u_0)|^2 \mathrm{d}x \right). \qquad (27)$$

From the inequalities (26) and (27) we get

$$\int\limits_{l_\beta} (u_\epsilon - u_0)^2 \mathrm{d}s \leqslant C_{19}\beta, \qquad (28)$$

since $\int\limits_{\Omega} |\nabla(u_\epsilon - u_0)|^2 \mathrm{d}x$ is uniformly bounded with respect to ϵ and because of Lemma 3.

By the imbedding theorem (7) for $(u_\epsilon - u_0)$ we obtain the estimate

$$\int\limits_{\Omega \backslash \Omega_\beta} (u_\epsilon - u_0)^2 \mathrm{d}x$$

$$\leqslant C_{20} \left(\int\limits_{\Omega \backslash \Omega_\beta} |\nabla(u_\epsilon - u_0)|^2 \mathrm{d}x + \int\limits_{l_\beta} (u_\epsilon - u_0)^2 \mathrm{d}s \right) \qquad (29)$$

with the constant C_{20}, which doesn't depend on ϵ, because of smoothness of $\partial\Omega$. From the estimates (28) and (29) we obtain

$$\int\limits_{\Omega} (u_\epsilon - u_0)^2 \mathrm{d}x$$

$$\leqslant C_{21} \left(\int\limits_{\Omega \backslash \Omega_\beta} |\nabla(u_\epsilon - u)|^2 \mathrm{d}x + \beta \right) + \int\limits_{\Omega_\beta} (u_\epsilon - u_0)^2 \mathrm{d}x.$$

By using the estimates (24) and (25), Theorems 1 and 2, we get (22) and (23), respectively. The theorem is proved.

We study also the limit behavior of the spectrum of the problem (1)–(3) as $\epsilon \to 0$. The question the behavior of spectrum of a boundary-value problem, when the boundary conditions are perturbed, was considered in [16]. The case when the sets Γ_ϵ^k are situated in a periodic way, was studied in [3]. In these lectures using the theorem on the limit behavior of the spectrum of the abstract operator sequence, which is proved in [10] (see also [17]), we study a nonperiodic case.

Consider the spectral problems which correspond to the boundary-value problems (1)–(3) and (10),(11):

$$L(u_\epsilon^k) + \lambda_\epsilon^k u_\epsilon^k = 0 \text{ in } \Omega, \tag{30}$$

$$u_\epsilon^k = 0 \text{ on } \gamma_\epsilon, \tag{31}$$

$$\frac{\partial u_\epsilon^k}{\partial \mu} + a(x)u_\epsilon^k = 0 \text{ on } \Gamma_\epsilon, k = 1,2,\ldots, \tag{32}$$

$$L(u_0^k) + \lambda_0^k u_0^k = 0 \text{ in } \Omega, \tag{33}$$

$$u_0^k = 0 \text{ on } \partial\Omega, k = 1,2,\ldots. \tag{34}$$

Here $u_\epsilon^k \in H^1(\Omega, \gamma_\epsilon)$, $u_0^k \in H^1(\Omega, \partial\Omega)$, $k = 1,2,\ldots$, the sets $\{\lambda_\epsilon^k\}$, $\{\lambda_0^k\}$, $k = 1,2,\ldots$, are eigenvalues such that $\lambda_\epsilon^1 \leqslant \lambda_\epsilon^2 \leqslant \cdots \leqslant \lambda_\epsilon^k \leqslant \cdots, \lambda_0^1 \leqslant \lambda_0^2 \leqslant \cdots \leqslant \lambda_0^k \leqslant \cdots$, the eigenvalues are repeated according to their multiplicities.

Define the operator $\mathcal{A}_\epsilon : L_2(\Omega) \to H^1(\Omega, \gamma_\epsilon)$, setting $\mathcal{A}_\epsilon f = -u_\epsilon$, where u_ϵ is the solution of the problem (1)–(3). The operator $\mathcal{A}_0 : L_2(\Omega) \to H^1(\Omega, \partial\Omega)$ is defined by the formula $\mathcal{A}_0 f = -u_0$, where u_0 is the solution of the problem (10),(11). Let $H_\epsilon = H_0 = L_2(\Omega)$, $V = H^1(\Omega, \partial\Omega)$ and let $\mathcal{R}_\epsilon$ be an identical operator in $L_2(\Omega)$.

Let us verify the conditions of the theorem from [10] (see Ch III, Theorem 1.4 and also section 1 of Chapter 4 of these lecture notes.) The condition C_1 is fulfilled automatically. It is easy to establish the positiveness, selfadjointness and compactness of the operators $\mathcal{A}_\epsilon, \mathcal{A}_0$. The norms $\|\mathcal{A}_\epsilon\|_{\mathcal{L}(H_\epsilon)}$ are uniformly bounded with respect to ϵ by virtue of Lemma 3.

Condition C_3 holds by Theorem 3. The sequence $\{\mathcal{A}_\epsilon f_\epsilon\}$ is bounded in $H^1(\Omega, \gamma_\epsilon)$ and, therefore, is compact in $L_2(\Omega)$ for any sequence $\{f_\epsilon\}$ bounded in $L_2(\Omega)$. Hence the condition C_4 is fulfilled.

Consider the spectral problems

$$\mathcal{A}_\epsilon u_\epsilon^k = \mu_\epsilon^k u_\epsilon^k, \quad \mu_\epsilon^1 \geqslant \mu_\epsilon^2 \geqslant \cdots, k = 1,2,\ldots,$$
$$\mathcal{A}_0 u_0^k = \mu_0^k u_0^k, \quad \mu_0^1 \geqslant \mu_0^2 \geqslant \cdots, k = 1,2,\ldots.$$

It is obvious that $\mu_\epsilon^k = (\lambda_\epsilon^k)^{-1}$, $\mu_0^k = (\lambda_0^k)^{-1}$. From Theorem 1.4, Ch III of

[10]
$$|\mu_\epsilon^k - \mu_0^k| \leqslant C_{22} \sup_{\substack{u \in N(\mu_0^k, \mathcal{A}_0) \\ \|u\|_{H_0}=1}} \|\mathcal{A}_\epsilon u - \mathcal{A}_0 u\|_{H_\epsilon},$$

$k = 1, 2, \ldots$, where $N(\mu_0^k, \mathcal{A}_0) = \{u \in H_0, \mathcal{A}_0 u = \mu_0^k u\}$. Thus the following theorem is proved:

Theorem 4. *There exist constants C_{23} and C_{24}, which don't depend on ϵ and such that for eigenvalues λ_ϵ^k and λ_0^k of problems (30)–(32) and (33),(34) respectively, the estimate $\left|\frac{1}{\lambda_\epsilon^k} - \frac{1}{\lambda_0^k}\right| \leqslant C_{23}|\ln \epsilon|^{-\delta}$ for sufficiently small ϵ is valid, where $0 < \delta < 2 - \frac{2}{n}$ and $N_\epsilon = O(|\ln \epsilon|^{(1-\frac{\delta}{2})n-1})$ as $\epsilon \to 0$. If the conditions of Theorem 2 are satisfied, then $\left|\frac{1}{\lambda_\epsilon^k} - \frac{1}{\lambda_0^k}\right| \leqslant C_{24}\epsilon^\alpha$ for sufficiently small ϵ, where $0 < \delta < n - 2$, $\alpha = \min(\delta, 1/(1+\sigma))$, $N_\epsilon = O(|\ln \epsilon|^2 \epsilon^{2-n+\delta})$ as $\epsilon \to 0$. The constants C_{23}, C_{24} do not depend on ϵ.*

The theorem which is analogous to Theorem 1 is proved for stationary linear elasticity, and the estimates which are similar to (12) are obtained. Also we considered the elliptic equations of the form (1) and the stationary linear elasticity system in a perforated domain Ω (see [18]). Let $\Omega_\epsilon = \Omega \backslash \{\bigcup_k \omega_k\}$, where the domain ω_k has a diameter ϵ, and we consider the equation in Ω_ϵ with the boundary conditions (2),(3) on $\partial\Omega$ and the Dirichlet boundary condition on $\partial\omega_k$. Then the theorems which are similar to Theorems 1–4 are valid. Moreover, we considered the problem when the Dirichlet condition is given on the boundary of some domains ω_k and a condition of the form (3) is given on the boundary of the other ω_k. In addition we suppose in this case, that the function $u \in H^1(\Omega_\epsilon)$ can be extended in $H^1(\Omega)$ in such a way that $\|u\|_{H^1(\Omega)} \leqslant C_{25}\|u\|_{H^1(\Omega_\epsilon)}$, where the constant C_{25} does not depend on ϵ.

Similar results have been proved in the case where we set some other type of boundary condition on γ_ϵ. The results of this section were obtained jointly with G.A. Chechkin.

References

[1] G.A. Chechkin, On boundary-value problems for a second order elliptic equation with oscillating boundary conditions. In: Nonclassical partial differential equations, Institute for Mathematics of the Siberian division of the Russian Academy of Sciences Press, Novosibirsk, 1988, 95–104.

[2] M. Lobo, E. Perez, Asymptotic behavior of an elastic body with a surface having small stack regions, *MMAN* **22**(4), 1988, 609–624.

[3] G.A. Chechkin, On the asymptotic properties of a partially fastened membrane, *Russian Math. Surveys* **44**, 1989, No 4, (*YMH*, **44**, No 4, 227).

[4] A. Brillard, M. Lobo, E. Perez, Homogénéisation des frontières par épi-convergence en élasticité lineaire, *MMAN*, **24**(1), 1990, 5–26.

[5] G.A. Chechkin, Homogenization of boundary-value problems with singular perturbation of the boundary conditions, *Mat. Sbornik* **184**(4), 1993, 99–150.

[6] A. Damlamian, Li Ta-Tsien (Li Daqian), Homogénéisation sur le bord pour des problèmes elliptiques, *C.R. Acad. Sci. Paris* **299**, 17, série 1, 1984, 859–862.

[7] A. Damlamian, Li Ta-Tsien (Li Daqian), Boundary homogenization for elliptic problems, *J.Math.Pure et Appl.* **66**, 1987, 351–361.

[8] A. Brillard, M. Lobo, E. Perez, Un probleme d'homogénéisation de frontiere en élasticité lineaire pour un corps cylindrique, *C.R. Acad. Sci. Paris* **311**, Ser. II, 1990, 15–20.

[9] K. Yosida, Functional analysis, Springer Verlag, 1965.

[10] O.A. Oleinik, A.S. Shamaev, G.A. Yosifian, Mathematical problems in elasticity and homogenization, North-Holland, Amsterdam, 1992.

[11] S.L. Sobolev, Some applications of functional analysis in mathematical Physics, Nauka, Moscow, 1988, AMS translation 1991.

[12] D. Gilbarg, N.S. Trudinger, Elliptic partial differential equations of second order, Springer Verlag, New York, 1977.

[13] G. Fichera, Alcuni recenti sviluppi della teoria dei problemi al contorno per le equazioni alle derivate parziali lineari, *Atti del Convegno Internationale sulle equazioni alle derivate parziali*, 1954, Trieste.

[14] G. Fichera, On a unified theory of boundary value problems for elliptic-parabolic equations of second order, *Boundary problems in differential equations*, University of Wisconsin Press, Madison, Wis., 1960, 97–120.

[15] M. Chicco, Principio di massimo per soluzioni di problemi al contorno misti per equazioni ellittiche di tipo variazionale, *Bull. Un. Math. Ital.* **4**, 1970, No 3, 384–394.

[16] A.A. Samarsky, On an influence of fastening to the eigen-frequencies of closed domains, *Uspechi Mat. Nauk* **5**, 1950, 3, 133–134.

[17] O.A. Oleinik, A.S. Shamaev, G.A. Yosifian, On a limit behavior of the spectrum of a sequence of operators which are defined in different Hilbert spaces, *Russian Math. Surveys* **44**, 1989, No 3.

[18] O.A. Oleinik, G.A. Chechkin, On boundary-value problems for the elasticity theory system with a rapidly changing type of boundary conditions, *Russian Math. Survey* **49**, 1994, No 4.

[19] O.A. Oleinik, G.A. Chechkin, On asymptotics of solutions and eigenvalues of the boundary-value problem with rapidly alternating boundary conditions for the system of elasticity, *Rendiconti Lincei: Matematica e Applicazioni*, 1996.

Index

Printed in the United States
By Bookmasters